Computer-Aided Design of Antimicrobial Lipopeptides as Prospective Drug Candidates

Computer-Aided Design of Antimicrobial Lipopeptides as Prospective Drug Candidates

Satya eswari Jujjavarapu, Swasti Dhagat
and Manisha Yadav

CRC Press
Taylor & Francis Group
Boca Raton London New York

CRC Press is an imprint of the
Taylor & Francis Group, an **informa** business

CRC Press
Taylor & Francis Group
6000 Broken Sound Parkway NW, Suite 300
Boca Raton, FL 33487-2742

First issued in paperback 2021

ISBN 13: 978-1-03-223877-7 (pbk)
ISBN 13: 978-1-138-49750-4 (hbk)

DOI: 10.1201/9781351018302

Library of Congress Cataloging-in-Publication Data

Names: Jujjavarapu, Satya eswari, author.
Title: Computer-aided design of antimicrobial lipopeptides as prospective drug candidates / Satya eswari Jujjavarapu, Swasti Dhagat, Manisha Yadav.
Description: Boca Raton, Florida : CRC Press, 2019. | Includes bibliographical references and index. | Summary: "Increase in antibiotic resistance has forced researchers to develop new drugs against microorganisms. Lipopeptides are produced as secondary metabolites by some microorganisms. Computer-aided Design of Antimicrobial Lipopeptides as Prospective Drug Candidates provides the identification of novel ligands for different antimicrobial lipopeptides. Along with identification, it also provides some of the in silico drug design processes, namely homology modeling, molecular docking, QSAR studies, drug ADMET studies and pharmacophore studies to check the ligand-lipopeptide interaction. Some lipopeptides have shown anti-cancerous properties too, and this book discusses the required templates to design new drugs using computational techniques"-- Provided by publisher.
Identifiers: LCCN 2019028794 | ISBN 9781138497504 (hardback) | ISBN 9781351018302 (ebook)
Subjects: LCSH: Peptide antibiotics--Design. | Drugs--Computer-aided design. | Drug development--Computer-aided design.
Classification: LCC RS431.P37 E 2019 | DDC 615.10285--dc23
LC record available at https://lccn.loc.gov/2019028794

Visit the Taylor & Francis website at
http://www.taylorandfrancis.com

and the CRC Press website at
http://www.crcpress.com

Contents

Preface

L IPOPEPTIDES ARE PRODUCED AS secondary metabolites from various microorganisms. They are amphiphilic molecules that contain a lipid moiety attached to a cyclic peptide. Such molecules possess surface-active properties, hence they are found to exhibit broad-spectrum potential antimicrobial properties. Due to the increase in antibiotic-resistant species of microorganisms, current antibiotics have been rendered ineffective, resulting in the demand for the development of novel and effective drugs to deal with these pathogens. Testing of different molecules to search for new drugs leads to the usage of various substrates and reagents. The identification of ligands is also a time-consuming process and the use of animal models for *in vivo* testing involves ethical concerns. This makes the process tedious and increases the overall expenditure of novel drug development. *In silico* modeling for drug design overcomes these disadvantages.

Computer-aided drug design uses *in silico* models and computational simulations to study the effect of various drug molecules on their target protein and facilitates the visualization of the underlying phenomenon of drug–ligand interaction. Drug design strategies are classified as structure-based and ligand-based. In structure-based drug design, the three-dimensional structure of the target protein is known, which helps in the identification of potential drug molecules, whereas in ligand-based drug design, knowledge of the structure of the drug molecule that binds to the corresponding target protein is required. In homology modeling, a three-dimensional structure of the target protein is generated based on the similarity with related protein structures. A molecular docking tool facilitates study of the interaction between a drug molecule and its target. *In silico* drug design also considers drug activity within the target biological system by studying its effect on the biological system and studying the properties of absorption, distribution, metabolism, excretion, and toxicity (ADMET). This book examines the properties of various lipopeptides that are produced as metabolites by many species of bacteria and have shown to possess antimicrobial properties. Thus, it is useful for students and researchers working in the field of drug design and discovery. This book also provides a step-by-step method for identifying drug molecules for the target protein and testing various properties for drug efficacy. Hence, it will be very helpful for beginners in this area.

Acknowledgments

First, our greatest regards to the almighty GOD for bestowing upon us the courage, unfailing source of support, comfort, and strength to complete this book successfully. We would like to express our gratitude to the Chhattisgarh Council of Science and Technology (CCOST) (Project number 2487/CCOST/MRP/2016, Raipur dated 25.01.2016), India, without which the conceived idea would have not been implemented successfully.

We are grateful to the Director of the National Institute of Technology, Raipur, and the Head of Department, Department of Biotechnology, National Institute of Technology, Raipur, for their continuous and unrelenting support.

We would also like to extend our gratitude to all the faculty members of Department of Biotechnology, National Institute of Technology Raipur, our colleagues, mentors, friends and family members for their emotional support.

Satya eswari Jujjavarapu: I would like to thank my brother for motivating me and making me successful in every aspect of life.

Authors

Satya eswari Jujjavarapu is currently working as an assistant professor in the Department of Biotechnology at the National Institute of Technology (NIT), Raipur, India. Her fields of specialization include bioinformatics, biotechnology, process modeling, evolutionary optimization, and artificial intelligence. She has more than 35 publications in SCI/Scopus indexed journals and 35 proceedings in international and national conferences. Her research contributions have received wide global citation. She has also published 6 book chapters and 4 books (currently in press) with international publishers. She is an active member of various organizations and has received various awards.

Swasti Dhagat is a research scholar in the Department of Biotechnology at the National Institute of Technology, Raipur, India. She has five research publications in peer-reviewed journals and a conference proceeding in an international conference in the field of *in silico* drug design of lipopeptides.

Manisha Yadav is a research scholar in the Department of Biotechnology at the National Institute of Technology, Raipur, India. Her area of specialization is in the field of bioinformatics with expertise in various computational tools and genomic software, proteomics, and drug design and discovery. She has SCI/Scopus publications in international journals and several international conference/workshop proceedings relevant to *in silico* drug design. Her research work centers on the computer-aided drug design of anticancer molecules and antibiotics for multi-drug resistant microbial strains.

CHAPTER 1

Lipopeptides and Computer-Aided Drug Design

1.1 WHAT ARE LIPOPEPTIDES?

Lipopeptide molecules consist of a lipid and a peptide connected to each other. These molecules are produced by bacteria and can be self-assembled into various structures (Hamley 2015, Kirkham et al. 2016). Lipopeptides are generally toll-like receptor agonists but some are used as antibiotics and certain works (Hamley et al. 2014) demonstrate their strong hemolytic and antifungal activity (Maget-Dana and Peypoux 1994). The interaction of lipopeptides with plasma membrane demonstrates their antibiotic properties (Nasir et al. 2013). Sterol components present on the plasma membrane participate in such interaction (Nasir and Besson 2011, Nasir and Besson 2012). Several bacterial species produce small molecules of peptide linked with lipid moiety, which are called lipopeptides.

Microbial lipopeptides: Lipopeptides derived from microorganisms as secondary metabolites are called microbial lipopeptides, which consist of antibiotic properties against various bacterial and fungal pathogens. Screening and identification of various novel lipopeptides are predominantly established from *Bacillus* species because of their prospective therapeutic applications for the purpose of human welfare. The ongoing research for the identification of novel antibiotics from bacterial lipopeptides aims to overcome the hazardous diseases associated with humans, plants and animals (Meena et al. 2017). Various categories of lipopeptides have been studied extensively in the context of biological control because of their effective antagonistic activities against the deadly pathogens. The weight of microbial lipopeptides is around 1.0–1.2 kDa, which exerts certain biochemical and physicochemical properties to fit into the category of antibiotics. Lipopeptides are considered as potent alternatives to combat against disease-causing pathogens that have acquired resistance to conventional antibiotics. Such pathogens are a major cause of life-threatening diseases. Apart from the antibiotic properties, lipopeptides consist of antiviral, anti-parasitic, anti-thrombolytic, hemolytic and anticancer properties. As an antitumor agent, lipopeptides are found with a potential to trigger programmed cell death or apoptosis in the malignant cells. The features of bacterial lipopeptides such as high biodegradability, low irritancy, lower toxicity and good compatibility for humans and animals make them suitable to be used as potent antibiotics (Meena et al. 2017). The biological activities and functional properties have increased their use in various domains and attracted scientists to produce lipopeptides. A strain of microorganism can produce several isoforms of lipopeptides, for example production of iturin, surfactin and fengycin is reported from *Bacillus subtilis* JKK328 by Yoon et al. (2005), and Xia et al (2014). reported their production from *Pseudomonas* species WJ6. The coproduction of lipopeptides surfactin, bacillomycin and plipastatin are reported from halophilic strain of *B. subtilis* BBK-1 (Roongsawang et al. 2002), and surfactin and plipastatin are coproduced by *Bacillus licheniformis* F2.2 (Thaniyavarn et al. 2003). Lipopeptides related to *Bacillus* (Ongena and Jacques 2008) and *Pseudomonas* (Raaijmakers et al. 2006) species are widely studied.

1.2 ADVANTAGES AND APPLICATIONS OF LIPOPEPTIDES

Bioactive chemical compounds are produced in wide variety from microorganisms during their growth phase (Biniarz et al. 2017). The surging demand of lipopeptides by leaps has no boundaries because of their extensive utilization for human well-being (Meena et al. 2016). Microorganisms coproduce lipopeptides with broader biopharmaceutical and biotechnological applications (Hsieh et al. 2008).

1.2.1 Biomedical and Therapeutic Applications of Lipopeptides

Amid the diversified category of biosurfactants, lipopeptides are taken into interest because they possess high surface activity and potential antibiotic activity against an array of pathogenic microorganisms. Polymyxin, the first lipopeptide, was isolated and discovered from *Bacillus polymyxa* (a soil bacterium) in 1949. Daptomycin (CubicinR) is the first cyclic lipopeptide antibiotic to receive approval from the Food and Drug Administration (FDA) in the United States. It is used to treat serious skin and blood infections caused by certain Gram-positive bacteria (Nakhate et al. 2013). Members of the *Bacillus* species are considered dedicated microbial factories because of the production of bioactive lipopeptides in large scale (Roongsawang et al. 2011, Wang et al. 2015, Meena et al. 2016).

1.2.2 Cyclic Lipopeptides: Potent Mosquito Larvicidal Agents

Mosquitoes, being blood-consuming insects, facilitate as a vector to spread various human diseases such as malaria, encephalitis, yellow fever, lymphatic filariasis, West Nile fever, and the like. The cell-free culture broth of the surfactin-producing strain of *B. subtilis* is capable of killing mosquito species such as *Anopheles stephensi*, *Culex quinquefasciatus* and *Aedes aegypti* at pupal and larval stages. As certain insecticides and biocontrol agents work effectively against mosquito larvae and pupae, the use of lipopeptides could be a good resource for application in control programs against malaria (Geetha et al. 2010). Public awareness is increasing because of human and environmental risks related to the use of chemical insecticides and pesticides. The pesticide-resistant population of insects is emerging as a new challenge besides the increasing prices of chemical pesticides. This has encouraged the hunt for new tools for vector control in an eco-friendly manner (Mittal 2003). In this context, the testing of various biological control agents for evaluating their potential to manage mosquito vectors is in progress in India and across many parts of the world. Toxins derived from some bacterial strains such as *B. sphaericues* (Bs), *B. thuringiensis* var. *israelensis* (Bti) are efficient in killing larvae of mosquito to a great extent even at a low dose. These are safe for non-target organisms too. The key strategy is using a mix of toxins for delaying the mosquitocidal proteins, which act upon various target sites within the insects (Giersch et al. 2009).

1.2.3 Antiparasitic Activity of Lipopeptides

Surfactin is a lipopeptide studied extensively for its antiparasitic and antiviral properties. Surfactin is a potent molecule with the capability of reducing the development of parasitosis. This acts either via directly exposing the molecule to spores or through incorporating into the luminal of midgut of bee (Porrini et al. 2010). The functioning of surfactin depends upon its uncompetitive inhibition for acylated peptides and the competitive inhibition for NAD+. Surfactin is a potent inhibitor of *in vitro* intra-erythrocytic growth of *Plasmodium falciparum* (Chakrabarty et al. 2008). Surfactin is found to be an alternative option for the treatment of nosemosis. The exposure of surfactin resulted in reduced infectivity of *Nosema ceranae* spores, which are the disease-causing agent of parasitic infection in *Apis mellifera* (Porrini et al. 2010). Nosemosis is a worldwide disease and *Nosema ceranae* is one of the known etiologic agents for the disease (Giersch et al. 2009). Furthermore, the introduction and administration of surfactin in the digestive system of a bee causes the reduced development of parasitoids (Porrini et al. 2010).

1.2.4 Antiviral Activity of Lipopeptides

Lipopeptide surfactin acts against various viruses like HSV-1 and HSV-2 (herpes simplex virus), semliki forest virus, vesicular stomatitis virus, feline calicivirus, murine encephalomyocarditis virus and simian immunodeficiency virus (Meena et al. 2016). The carbon chain length of cyclic lipopeptides surfactin has influential capability for the deactivation of viruses (Singla et al. 2014). Surfactin significantly deactivates the infection of enveloped viruses, especially retrovirus and herpes virus, rather than acting upon non-enveloped viruses. This antiviral activity of lipopeptides is indicated by a membrane-active surfactant property and their physicochemical interaction with lipid membrane of virus. The carbon atom number present in the acyl chain of surfactin also plays a vital role in the inactivation of virus. The capacity for inactivation of virus rises with the increasing hydrophobicity of fatty acid. The permeation of surfactin by lipid bilayer takes place during the deactivation of virus, thereby inducing the breakdown of the envelope completely and also the viral proteins participating in the adsorption and penetration of the virus at the target cells. The absence of viral proteins accounts for the failure of viral infectivity. Thus, lipopeptides demonstrate the antiviral activity (Sachdev and Cameotra 2013).

1.2.5 Antitumor Activity and Lipopeptides-Induced Apoptotic Pathway

Lipopeptides induce oxidative stress, which leads to reactive oxygen species (ROS) production in cancer cells treated by lipopeptides. Subsequently, the oxidative stress induces apoptosis in the cells, as indicated by the fragmentation and condensation of nuclei. An additional marker is DNA nicking, which is evidence of apoptotic cell death and can be notified by FACS-based tunnel assay. The concentration of lipopeptides influences the extensive DNA nicking (Alonso et al. 2003).

An example of potent lipopeptide is surfactin, which is considered a very versatile bioactive molecule and also exhibits antitumor activity (Sachdev and Cameotra 2013). It is reported that surfactin acts as a cytotoxic agent against proliferating human colon carcinoma cell lines like HT29 and HCT-15 (Sivapathasekaran et al. 2010). The growth of transformed cells is inhibited by surfactin because of cell cycle arrest and the introduction of the apoptotic pathway by suppressing regulating signals (i.e. ERK and Akt) required for cell survival (Kim et al. 2007). It is noted that increased concentration of surfactin and higher exposure time lead to the decreased percentage of viable cells, which is an indication of cytotoxic or cytostatic effect against cell lines of breast cancer (Duarte et al. 2014). According to a study, proliferation inhibition and apoptosis induction by surfactin in a dose-dependent manner is mediated through a ROS/JNK mediated mitochondrial/caspase pathway (Lee et al. 2012). Surfactin is capable of generating ROS. These are the key regulators in apoptotic pathway (Cao et al. 2010). The strategy of lipopeptides-induced apoptotic cell death has emerged as a preventative measure for cancer treatment.

1.2.6 Anti-Obesity Activity of Lipopeptides

Obesity is considered to be a lifestyle disorder specifically in developing countries. It is prevailing globally and pertains to unbalanced food habits, which include consumption of fast food, sweet products containing high fructose corn syrup and a lifestyle led by reduced physical activity (Bray 2013). Inhibitory activity of pancreatic lipase is used to explore the potency of natural products (Lunagariya et al. 2014). As a unique class of biosurfactant, lipopeptide, has emerged as a promising molecule owing to its structural novelty, diverse properties and versatility, which makes it suitable for use in advanced therapeutic applications (Gudiña et al. 2013). Lipopeptides from *B. subtilis* SBP1 can be a suitable drug for treating majorly metabolic disorders related to obesity. Oral administration of *B. subtilis* lipopeptides can be used for effective management of body weight. Crude lipopeptide of *B. subtilis* SBP1 possesses both curative and protective action on obese people and has been proven to reduce body weight of obese rats by reducing the activity of serum pancreatic lipase (Zouari et al. 2016). Thus, it appears suitable to treat hyperlipidemia without apparent side effects.

1.2.7 Thrombolytic Activity of Lipopeptides

The plasminogen-plasmin system involved in dissolving blood clots in various physiological and pathological processes is required in proteolysis. Proteolytic activation of zymogen plasminogen is performed via tissue type and urokinase type plasminogen activator (Singla et al. 2014). Activation of prourokinase and plasminogen is an important mechanism to initiate and propagate fibrinolytic activity. According to the reports available, an extensively studied lipopeptide, surfactin, at the concentration of 3–20 µmol/L leads to enhanced activation of prourokinase and possesses the ability to change the conformation of plasminogen, which increases the fibrinolysis further *in vitro* and *in vivo* (Kikuchi and Hasumi 2002). Experimental studies on a mouse pulmonary embolism model showed that injection of surfactin C in combination with prourokinase was able to increase lysis of the plasma clot. Surfactin also possesses the ability to prevent aggregation of platelets, which inhibits the formation of fibrin clots (Blanchard and Montagnier 1994). Lipopeptide-based drugs could also increase fibrinolysis by facilitating dispersal of fibrinolytic agents (Meena et al. 2017). The detergent-like property of lipopeptides is involved in the antiplatelet activity and it acts on downstream signaling pathways (Kim et al. 2006). Furthermore, lipopeptides are advantageous over other thrombolytic agents due to fewer associated side effects and to being potential drug candidates for long-term use as clot bursting agents.

1.3 COMPUTER-AIDED DRUG DESIGNING (*IN SILICO* DRUG DESIGN)

In silico drug design is a computational approach of designing or discovering a new drug molecule. This is considered as a theoretical approach with computational assistance, which shortens the path of research in finding new drugs. Drug design refers to a rational design or simply rational drug design. It is the innovative process based upon the information of a biological target for finding new medicines (Liljefors et al. 2002). In the current era of computer-aided drug design or *in silico* drug discovery, the research focuses on understanding the disease mechanism, which is followed by discovery of lead compound and target identification. The drugs are commonly small organic molecules that participate in activation and inhibition of biomolecules such as proteins, which results in facilitating the therapeutic advantage to patients. Drug design deals with a rudimentary designing of molecules that is complementary in terms of charge and molecular shape to the respective biomolecular target to which they can interact and bind. *In silico* drug design relies on techniques of computer modeling (Merz et al. 2010) and such modeling is called computer-aided drug design (see Figure 1.1). Drug design that deals with a known three-dimensional structure of target protein is referred to as structure-based drug design (Merz et al. 2010). The current trend of cost-effective outcomes of public health and personalized medicine systems is based upon molecular states, i.e. DNA to RNA and protein, which is the fundamental concept of drug discovery (Barabási et al. 2011, Chen and Butte 2013). Molecular characterization is an important step in building up a system of drug design, while environmental and bodily influences also need to be considered (Culligan et al. 2014, Garber 2015). However, measures for safety and regulatory requirements have to be taken into account while going with the *in silico* approach of drug design (Wang and Xie 2014, Katsila et al. 2016).

1.3.1 Homology Modeling (HM)

Homology modeling (HM) is also referred to as comparative modeling of proteins. This method involves the prediction of an atomic resolution model of a target protein from its primary sequence of amino acids. The prediction of the 3D structure of the "target" protein sequence is

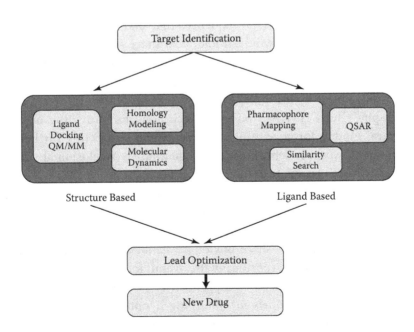

FIGURE 1.1　Graphical depiction of computer-aided drug discovery.

performed using a known experimental 3D structure of homologous protein which is used as a "template" for the target sequence. HM is based upon the recognition of one or more similar structures of known protein that resembles the probable structure of query sequence. By producing an alignment, it maps the residues of query sequence to the residues of template sequence. This is based on the concept of conserved protein structures among the homologous protein sequences. The sequences with less than 20% sequence similarity will have notably different structures (Chothia and Lesk 1986). Naturally occurring similar sequences and evolutionary-related homologous sequences possess similar protein structures. It is found that protein 3D structures are evolutionarily more conserved than the expected conserved sequence alone (Kaczanowski and Zielenkiewicz 2010). Once the sequence alignment is performed, then the template structure is used to generate a homologous structural model of the target protein. The quality of the sequence alignment and similarity with the template structure affect the quality of the built homology model. This approach is complicated by the occurrence of alignment gaps (also called indels), which indicate those structural regions of target sequence that are not present in template sequence. Thus, the quality of a model declines with the decreased sequence similarity. However, HM is used to reach a qualitative conclusion concerned with the biochemistry of query or target sequence. The protein sequences that share detectable structure similarity, particularly the overall fold, also share some structural similarity and exhibit similar functions too. The basis of HM is that it is very time-consuming and difficult to acquire the experimental structures of protein of interest by using methods like protein NMR and X-ray crystallography. Thus, HM is a powerful technique for providing obvious structural models for generating hypotheses and redirecting further for experimental work. It helps in getting a qualitative conclusion about the biochemistry of a template or query sequence.

Motive of HM: The motive of HM method is the observation that the tertiary structure of protein is more conserved than the primary sequence of amino acid (Martí-Renom et al. 2000). Hence, proteins that have diversified sequences but still share obvious resemblance will share similar structural properties too, specifically the overall fold. Since it is time-consuming and difficult to acquire experimental structures, HM facilitates generating structural models for obtaining hypothetical data about the functions of protein. These theoretical structures further provide directions for experimental work.

Steps in HM

1. Sequence alignment and template selection: The identification of suitable template protein structure is the first and most critical step in HM. The template identification method is simply relying on sequential pairwise sequence alignment. This step is supported by a technique of similarity search using database mining such as BLAST and FASTA. Further sensitive and advanced methods are based upon multiple sequence alignment in which PSI-BLAST is a commonly used method. The method updates iteratively to a position-specific scoring matrix (PSSM) and successful identification of much distantly related homologous sequences.

2. Model generation: To acquire 3D structure with the identified template and alignment results, a 3D structural model of the target sequence that represents as a set of Cartesian coordinates for every atom of protein is generated (Baker and Sali 2001, Zhang 2008).

3. Fragment assembly: Originally, the method for HM was based upon the assembly of a whole model from conserved fragments of structure, which is identified from closely related solved structures. As an example, a modeling study of serine proteases in mammals found a sharp distinction involving "core" structural regions, which is conserved in all experimental structures of the class, and variable regions usually located in the loops where major parts of the sequence dissimilarities are confined. Hence, unsolved or unknown protein structures can be modeled by the construction of conserved core first and then by substitution of variable regions of other proteins in the solved structures set (Greer 1981).

4. Segment matching: The segment-matching method divides the target into a series of short segments, each of which is matched to its own template fitted from the Protein Data Bank. Thus, sequence alignment is done over segments rather than over the entire protein. Selection of the template for each segment is based on sequence similarity, comparisons of alpha carbon coordinates and predicted steric conflicts arising from the van der Waals radii of the divergent atoms between target and template (Levitt 1992).

5. Satisfaction of spatial restraints: Most commonly, the current method of HM is inspired by the requirement of calculations for

the construction of a 3D model from generated data through NMR spectroscopy. One of many templates – target alignment utilized for the construction of a geometrical criteria set is used to convert into probability density functions for every restraint. The application of restraints to the main internal coordinates of protein, dihedral angles and protein backbone distances are the basis for the procedure of global optimization. This is originally used to refine the positions iteratively for all heavy atoms of the protein through conjugate gradient energy minimization (Šali and Blundell 1993).

6. Loop modeling: Certain regions in the target sequence that are not aligned with the template sequence can be modeled through loop modeling. These are susceptible to modeling errors majorly and with high frequency occurrence while the template and target have low sequence similarity. The coordinates of mismatched segments predicted through loop modeling programs are comparatively less accurate than those obtained from simply copying the coordinates of a known structure, particularly if the loop is longer than ten residues.

7. Model assessment: Assessment of structures obtained by HM without reference to the original target structure is generally accomplished with two methods: physics-based energy calculations and statistical potentials. Both methods generally produce energy estimation (an energy-like analog) for models to be assessed. To determine the acceptable cutoffs, independent criteria are required. Both of the methods do not exceptionally correlate well with true structural accuracy, particularly on protein types diminished in the PDB, for instance membrane proteins.

The Radboud University Nijmegen "What If" software package provides "What Check" software, which is one of the options to obtain a very extensive model validation report. It extensively analyses to produce a document of many pages with nearly 200 administrative and scientific aspects of the model. What Check is available as a free server and can be used for the validation of experimentally determined macromolecule structures.

1.3.2 Molecular Docking Simulations (MDS)

In the arena of drug discovery, the molecular docking method is used to predict the favored orientation of two molecules binding to each other

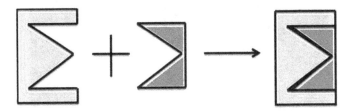

FIGURE 1.2   Protein and ligand interaction.

to form a stable complex (see Figure 1.2) (Lengauer and Rarey 1996). The prediction of binding affinity or the strength of association of two molecules is determined by using the scoring function if the predominant binding mode is known. Molecular docking is an important tool in computer-aided drug design and structural molecular biology. The aim of ligand – protein docking is the prediction of preferred orientation of a ligand with a protein molecule of known 3D conformations. Efficient methods of docking identify spaces of high dimension and correctly rank the candidate docking by using a scoring function. Molecular docking is generally used to rank the results and propose the structural hypothesis that the most important part of lead optimization is how a target is inhibited by ligand. Apart from that, virtual screening of compounds from large libraries can also be performed using docking. With the docking method, the setting up of the structure of input molecule is as significant as the docking itself (Morris and Lim-Wilby 2008). The illustration of the mechanism of docking between a ligand and a protein molecule is shown in Figures 1.3 and 1.4.

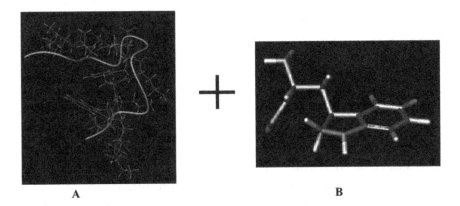

A   B

FIGURE 1.3   Lipopeptide structure (A) and ligand (B).

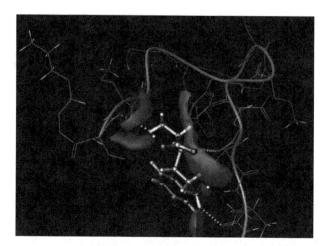

FIGURE 1.4 Docking of ligand and lipopeptides (depiction of drug–ligand interaction).

The association of biomolecules like lipids, proteins, carbohydrates and nucleic acid plays an important role in the process of signal transduction. In addition, the relative orientation of participating molecules in the interaction may have an impact on the type of signal produced such as antagonism vs. agonism. Furthermore, the ability to predict the binding conformation of small molecules to their appropriate binding site of target molecule is the basis of molecular docking. This is most likely to be used for structure-based drug design. To elucidate the fundamental biochemical processes and rationale of drug design, the characterization of binding behavior plays an indispensable role (Kitchen et al. 2004). Molecular docking can be comparted to a "lock and key" model, in which finding the correct relative orientation of one "key" can open the "lock" (a key hole present on the surface of the lock provides the direction to turn the key once it is inserted). Here, the protein can be assumed as a lock and the ligand can be assumed as a key. Moreover, protein and ligand are both flexible, so the analogy of "hand-in-glove" may be more suitable than "lock-and-key" (Jorgensen 1991). The protein and the ligand adjust their conformation to accomplish the overall "best-fit" during the course of docking and such conformational adjustment is referred to as "induced-fit," which results in overall binding (Wei et al. 2004). Molecular docking is based upon computational simulations for molecular recognition. This method aims at the optimization of conformations of ligand and protein in the best-fit orientation to bind together and the relative orientation of both for the interaction with the concerned protein in such a way that the free energy of the system is minimized (see Figure 1.4). Molecular docking is a computer-assisted approach, which is facilitated by various software available with versatile features. In this book, molecular docking is performed to understand the antagonistic properties of lipopeptides to inhibit the target molecule for exhibiting their antibiotic property. The study is done by using a very powerful software named Schrödinger that is loaded with various features and applications to be used in the steps of molecular docking and drug design. Schrödinger's Maestro visualization program (Schrödinger Release: Maestro – version 10.5, Schrödinger; LLC, NY 2016 – 1) was used.

1.3.3 Study of QSAR

QSAR modeling is a study of the quantitative structure activity relationship. The modeling contains molecular predictors, which consist of either theoretical data of molecular descriptors or physicochemical properties of a chemical molecule. The response variables of QSAR are based upon the biological activity of chemicals. QSAR models are first used to summarize the supposed relationship of chemical structures with their biological activity in chemical dataset. The second objective of using QSAR study is the prediction of the activity of a new chemical. Moreover, QSAR models are regression analyses and classification models. Regression models of QSAR work on the basis of relatedness of a set of "predictor" variable (X) and the potential of the response variable (Y), whereas the classical model of QSAR relates a categorical value of response variable and predictor variable. For example, the biological activity of a chemical compound can be expressed quantitatively as the substrate's concentration, which is required to exhibit some biological response. The purpose of QSAR models is based on the concept that if the structure or physicochemical properties are expressed in numbers, it is easy to establish a mathematical relation and the quantitative structure activity relationship between both of the molecules. If the mathematical expressions are validated carefully (Tropsha et al. 2003, Chirico and Gramatica 2012), they could be used for the prediction of modeled response of other chemical structures (Tropsha 2010). The significant approach of QSAR modeling is its ability to depict a correlation between the structures of a set of molecules and the target response. The general workflow of QSAR includes the collection of a set of inactive and active molecules against a target. It produces the descriptors that describe the physicochemical and structural properties. Thus, the model is used for correlating such descriptors with their experimental activities, which results as a predictive tool for new molecular entity. The algorithms of QSAR are evolving continuously and involve several 3D and 2D descriptors, which can be physicochemical or structural, such as molecular weight, rotatable bonds, volume, atom type, interatomic distances, electronegativity, molecular walk counts, aromaticity, atom distribution and salvation properties. As per the increasing complexity, these can be described at multiple levels.

- Mathematical model for QSAR: Activity = f (physiochemical properties/or structural properties) + error

- The error comprises of model error (biased) and observational variability, which is the variability in observations on a correct model also.

This book includes computational drug discovery of lipopeptides as proposed antimicrobial drugs. The discovery of a new drug includes the study of such models which depict the QSAR of the proposed drug with established drugs. This is also done in order to determine the quantity of chemicals required to exhibit its activity. QSAR study provides such mathematical regression analysis using software-based study to provide the information of molecular descriptors and predictors by comparing with the descriptors from data sets of compounds or established drug.

1.3.4 Pharmacokinetics/ADMET Study

ADMET study is a complete package for studying the fate of consumed drug molecule inside the body and the toxicity caused by the drug. This is a combined study of each level of metabolism starting from the absorption of a pharmaceutical compound into the blood stream, distribution at various sites where they need to show their activity, metabolism and excretion and also includes the toxicity caused by excreted drug molecule outside in the environment. It is generally performed to validate the drug at each level of metabolism within the specified limit. In pharmacology and pharmacokinetics, ADMET study is an abbreviation of absorption, distribution, metabolism, excretion and toxicity. Ultimately, the study is to understand the disposition of a drug molecule within an organism. The kinetics of drug exposure to the tissues and consequent effects of pharmacological activity and the performance of drug are influenced by the following four criteria.

1.3.4.1 Absorption/Administration (Pharmacokinetics)

A compound can reach a tissue, if it is taken into the bloodstream. Usually, a drug is administered often through mucous surfaces such as the digestive tract, i.e. intestinal absorption before it is taken up by the target cells. Factors like poor compound solubility, intestinal transit time, gastric emptying time, inability to permeate the intestinal wall and chemical instability in the stomach are responsible for reducing the extent of drug absorption after oral administration. Critically, absorption determines the bioavailability of a compound. Drugs with poor absorption are less desirable for oral administration, such as by inhalation or intravenously (e.g. zanamivir). The modes of drug administration are considered important.

1.3.4.2 Distribution (Pharmacology)/Dispersion or Dissemination of Substances

The compound should be carried to its site of action or effector site mostly through the bloodstream. The drug may dispense into muscles and organs from blood, usually to varying extents. Once the compound has entered into systemic circulation either via intravascular injection or through absorption from any of the different extracellular sites, it is subjected to several distribution levels that lead to lower plasma concentration. Distribution is a reversible transfer of a drug molecule between one compartment to another. There are various factors that affect drug distribution, which include molecular size, regional blood flow, polarity and binding to serum proteins and complex formation. Distribution can be a severe issue at some natural barriers such as the blood–brain barrier.

1.3.4.3 Metabolism

A drug molecule commences breaking down as it enters the body. The metabolism of small molecules of drug is mainly carried out in the liver via redox enzymes, which are termed as cytochrome P450 enzymes. In the process of metabolism, it deals with converting initial parent compound into new compound, i.e. metabolites. Metabolism leads to deactivation of the administered dose of parent drug, when metabolites are pharmacologically inactive which generally reduces the effect of the drug in the body. Metabolites may sometimes be pharmacologically more active than the parent drug. When so, they are called as prodrugs.

1.3.4.4 Excretion of the Drug

It is important to remove drug compounds and their metabolites from the body through excretion. This is usually carried out via the kidneys (urine) or feces. Foreign substances can accumulate in the body and can have adverse effects on normal metabolism if excretion is not completed. There are three main sites of drug excretion in the body. Kidneys are the most important and are involved in the excretion of products through urine. The process of fecal excretion or biliary excretion initiates in the liver and passes via gut until the products are excreted with feces or waste products. The last important method of excretion is via lungs such as with anesthetic gases.

The kidney involves three main mechanisms for the excretion of drugs:

1. Glomerular filtration of unbound drug.

2. Active secretion of drug (free or protein bound) by transporters (e.g. cations such as histamine, choline or anions such as penicillin, urate, sulphate conjugates, glucuronide).

3. Filtrate is 100-folds concentrated in tubules for a favorable concentration gradient, therefore, it may secrete through passive diffusion and pass out via urine.

1.3.4.5 Toxicity

The potential of toxicity of a drug compound is taken into account using ADMET Tox or ADMET study. Computational chemistry is used to predict the toxicity or toxic effects of compounds through methods such as QSAR or QSPR (quantitative structure property relationship). ADMET is influenced by the route of administration of drug (Balani et al. 2005, Singh 2006, Tetko et al. 2006).

1.3.5 Pharmacophore Properties

Pharmacophore study is the conceptual and theoretical description of molecular features that are essential for molecular recognition of a ligand by a biomolecule. The IUPAC (International Union of Pure and Applied Chemistry) defines pharmacophore study as an assembly of electronic and steric features which are essential for ensuring the most favorable interaction with a specific biological target to inhibit or trigger the biological response (Wermuth et al. 1998). The pharmacophore model is used to define how structurally dissimilar ligands interact with a common receptor site. Moreover, pharmacophore models are used for identification of novel ligands with binding affinity for same receptor through virtual screening or *de novo* design. Distinctive features of pharmacophore consist of aromatic rings, hydrophobic centroids, hydrogen bond donors and acceptors, anions and cations. Such projected pharmacophore points might be present in the receptor or may be presumed to be located on the ligand itself. To identify the novel ligands, features must match various chemical groups with similar or related properties. Receptor–ligand interactions are typically polar negative, polar positive or hydrophobic. A pharmacophore model is well defined with the incorporation of hydrogen bond vectors and hydrophobic volumes.

1.4 PHARMACOPHORE STUDY AS APPLICATION FOR DRUG-RELATED ACTIVITIES

Modern computational chemistry entertains pharmacophore study for defining the essential features of one or various molecules of similar biological activity. A diverse chemical compound database is searched for more molecules that exhibit identical features precisely arranged in the similar relative orientation. Pharmacophore models are generally the preliminary point for the development of 3D-QSAR models. The related concept of "privileged structure" and such tools are defined as molecular frameworks, which are able to provide required ligands for various types of enzyme targets or receptors with sensible structural modifications aiding to drug discovery (Duarte et al. 2007).

1.5 CONCLUSIONS

Lipopeptides appear to be antibiotics of a novel class which amazingly exhibit a broader category featuring surfactant, antibacterial, antiviral, antifungal, antimycoplasma, antilarval, antiparasitic, antitumor, anticancer and antithrombocytic activities. Among the genera of bacteria, the *Bacillus* species prominently produces extracellular lipopeptides. However, for the commercial utilization of lipopeptides as potent antifungal and antibacterial molecules, there is a need to produce them in sufficient quantity by the methods of cloning appropriate genes in the proficient expression vectors. Efficient targeting of cancer and tumor cells could be achieved by the conjugation of lipopeptides with tumor-specific surface-binding molecules. The compatibility with cell membrane, low molecular mass, biodegradability and eco-friendliness of lipopeptides put them in an interesting category of antibiotics.

REFERENCES

Alonso, M., Tamasdan, C., Miller, D. C. and Newcomb, E. W. 2003. Flavopiridol induces apoptosis in glioma cell lines independent of retinoblastoma and p53 tumor suppressor pathway alterations by a caspase-independent pathway1. *Molecular Cancer Therapeutics* 2: 139–150.

Baker, D. and Sali, A. 2001. Protein structure prediction and structural genomics. *Science* 294: 93–96.

Balani, S. K., Miwa, G. T., Gan, L.-S., Wu, J.-T. and Lee, F. W. 2005. Strategy of utilizing in vitro and in vivo ADME tools for lead optimization and drug candidate selection. *Current Topics in Medicinal Chemistry* 5: 1033–1038.

Barabási, A.-L., Gulbahce, N. and Loscalzo, J. 2011. Network medicine: a network-based approach to human disease. *Nature Reviews Genetics* 12: 56.

Biniarz, P., Łukaszewicz, M. and Janek, T. 2017. Screening concepts, characterization and structural analysis of microbial-derived bioactive lipopeptides: a review. *Critical Reviews in Biotechnology* 37: 393–410.

Blanchard, A. and Montagnier, L. 1994. AIDS-associated mycoplasmas. *Annual Review of Microbiology* 48: 687–712.

Bray, G. A. 2013. *Energy and Fructose from Beverages Sweetened with Sugar or High-Fructose Corn Syrup Pose a Health Risk for Some People*, Oxford University Press.

Cao, X.-H., Wang, A.-H., Wang, C.-L., et al. 2010. Surfactin induces apoptosis in human breast cancer MCF-7 cells through a ROS/JNK-mediated mitochondrial/caspase pathway. *Chemico-Biological Interactions* 183: 357–362.

Chakrabarty, S. P., Saikumari, Y. K., Bopanna, M. P. and Balaram, H. 2008. Biochemical characterization of *Plasmodium falciparum* Sir2, a NAD+-dependent deacetylase. *Molecular and Biochemical Parasitology* 158: 139–151.

Chen, B. and Butte, A. J. 2013. Network medicine in disease analysis and therapeutics. *Clinical Pharmacology & Therapeutics* 94: 627–629.

Chirico, N. and Gramatica, P. 2012. Real external predictivity of QSAR models. Part 2. New intercomparable thresholds for different validation criteria and the need for scatter plot inspection. *Journal of Chemical Information and Modeling* 52: 2044–2058.

Chothia, C. and Lesk, A. M. 1986. The relation between the divergence of sequence and structure in proteins. *The EMBO Journal* 5: 823–826.

Culligan, E. P., Sleator, R. D., Marchesi, J. R. and Hill, C. 2014. Metagenomics and novel gene discovery: promise and potential for novel therapeutics. *Virulence* 5: 399–412.

Duarte, C., Gudiña, E. J., Lima, C. F. and Rodrigues, L. R. 2014. Effects of biosurfactants on the viability and proliferation of human breast cancer cells. *AMB express* 4: 40.

Duarte, C. D., Barreiro, E. J. and Fraga, C. A. 2007. Privileged structures: a useful concept for the rational design of new lead drug candidates. *Mini Reviews in Medicinal Chemistry* 7: 1108–1119.

Garber, K. 2015. *Drugging the Gut Microbiome*, Nature Publishing Group.

Geetha, I., Manonmani, A. and Paily, K. 2010. Identification and characterization of a mosquito pupicidal metabolite of a *Bacillus subtilis* subsp. subtilis strain. *Applied Microbiology and Biotechnology* 86: 1737–1744.

Giersch, T., Berg, T., Galea, F. and Hornitzky, M. 2009. *Nosema ceranae* infects honey bees (*Apis mellifera*) and contaminates honey in Australia. *Apidologie* 40: 117–123.

Greer, J. 1981. Comparative model-building of the mammalian serine proteases. *Journal of Molecular Biology* 153: 1027–1042.

Gudiña, E. J., Rangarajan, V., Sen, R. and Rodrigues, L. R. 2013. Potential therapeutic applications of biosurfactants. *Trends in Pharmacological Sciences* 34: 667–675.

Hamley, I. W. 2015. Lipopeptides: from self-assembly to bioactivity. *Chemical Communications* 51: 8574–8583.

Hamley, I. W., Kirkham, S., Dehsorkhi, A., et al. 2014. Toll-like receptor agonist lipopeptides self-assemble into distinct nanostructures. *Chemical Communications* 50: 15948–15951.

Hsieh, F.-C., Lin, T.-C., Meng, M. and Kao, S.-S. 2008. Comparing methods for identifying *Bacillus* strains capable of producing the antifungal lipopeptide iturin A. *Current Microbiology* 56: 1–5.

Jorgensen, W. L. 1991. Rusting of the lock and key model for protein-ligand binding. *Science* 254: 954–956.

Kaczanowski, S. and Zielenkiewicz, P. 2010. Why similar protein sequences encode similar three-dimensional structures? *Theoretical Chemistry Accounts* 125: 643–650.

Katsila, T., Spyroulias, G. A., Patrinos, G. P. and Matsoukas, M.-T. 2016. Computational approaches in target identification and drug discovery. *Computational and Structural Biotechnology Journal* 14: 177–184.

Kikuchi, T. and Hasumi, K. 2002. Enhancement of plasminogen activation by surfactin C: augmentation of fibrinolysis in vitro and in vivo. *Biochimica et Biophysica Acta (BBA)-Protein Structure and Molecular Enzymology* 1596: 234–245.

Kim, S. D., Park, S. K., Cho, J. Y., et al. 2006. Surfactin C inhibits platelet aggregation. *Journal of Pharmacy and Pharmacology* 58: 867–870.

Kim, S.-Y., Kim, J. Y., Kim, S.-H., et al. 2007. Surfactin from *Bacillus subtilis* displays anti-proliferative effect via apoptosis induction, cell cycle arrest and survival signaling suppression. *FEBS Letters* 581: 865–871.

Kirkham, S., Castelletto, V., Hamley, I. W., et al. 2016. Self-assembly of the cyclic lipopeptide daptomycin: spherical micelle formation does not depend on the presence of calcium chloride. *ChemPhysChem* 17: 2118–2122.

Kitchen, D. B., Decornez, H., Furr, J. R. and Bajorath, J. 2004. Docking and scoring in virtual screening for drug discovery: methods and applications. *Nature Reviews Drug Discovery* 3: 935.

Lee, J. H., Nam, S. H., Seo, W. T., et al. 2012. The production of surfactin during the fermentation of cheonggukjang by potential probiotic *Bacillus subtilis* CSY191 and the resultant growth suppression of MCF-7 human breast cancer cells. *Food Chemistry* 131: 1347–1354.

Lengauer, T. and Rarey, M. 1996. Computational methods for biomolecular docking. *Current Opinion in Structural Biology* 6: 402–406.

Levitt, M. 1992. Accurate modeling of protein conformation by automatic segment matching. *Journal of Molecular Biology* 226: 507–533.

Liljefors, T., Krogsgaard-Larsen, P. and Madsen, U. 2002. *Textbook of Drug Design and Discovery*, CRC Press.

Lunagariya, N. A., Patel, N. K., Jagtap, S. C. and Bhutani, K. K. 2014. Inhibitors of pancreatic lipase: state of the art and clinical perspectives. *EXCLI Journal* 13: 897.

Maget-Dana, R. and Peypoux, F. 1994. Iturins, a special class of pore-forming lipopeptides: biological and physicochemical properties. *Toxicology* 87: 151–174.

Martí-Renom, M. A., Stuart, A. C., Fiser, A., et al. 2000. Comparative protein structure modeling of genes and genomes. *Annual Review of Biophysics and Biomolecular Structure* 29: 291–325.

Meena, K., Dhiman, R., Sharma, A. and Kanwar, S. 2016. Applications of lipopeptide (s) from a *Bacillus* sp: an overview. *Research Journal of Recent Sciences* 5: 50–54.

Meena, K., Sharma, A. and Kanwar, S. 2017. Microbial lipopeptides and their medical applications. *Annals of Pharmacology and Pharmaceutics* 2(24): 1126.

Merz Jr, K. M., Ringe, D. and Reynolds, C. H. 2010. *Drug Design: Structure-and Ligand-Based Approaches*, Cambridge University Press.

Mittal, P. 2003. Biolarvicides in vector control: challenges and prospects. *Journal of Vector Borne Diseases* 40: 20.

Morris, G. M. and Lim-Wilby, M. 2008. Molecular docking. In *Molecular Modeling of Proteins*, 365–382. Springer.

Nakhate, P., Yadav, V. and Pathak, A. 2013. A review on daptomycin; the first US-FDA approved. Lipopeptide antibiotics. *Journal of Scientific and Innovative Research* 2: 970–980.

Nasir, M. N. and Besson, F. 2011. Specific interactions of mycosubtilin with cholesterol-containing artificial membranes. *Langmuir* 27: 10785–10792.

Nasir, M. N. and Besson, F. 2012. Interactions of the antifungal mycosubtilin with ergosterol-containing interfacial monolayers. *Biochimica et Biophysica Acta (BBA)-Biomembranes* 1818: 1302–1308.

Nasir, M. N., Besson, F. and Deleu, M. 2013. Interactions des antibiotiques ituriniques avec la membrane plasmique. Apport des systèmes biomimétiques des membranes (synthèse bibliographique). *Biotechnologie, Agronomie, Société et Environnement* 17: 505–516.

Ongena, M. and Jacques, P. 2008. *Bacillus* lipopeptides: versatile weapons for plant disease biocontrol. *Trends in Microbiology* 16: 115–125.

Porrini, M. P., Audisio, M. C., Sabaté, D. C., et al. 2010. Effect of bacterial metabolites on microsporidian *Nosema ceranae* and on its host *Apis mellifera*. *Parasitology Research* 107: 381–388.

Raaijmakers, J. M., De Bruijn, I. and de Kock, M. J. 2006. Cyclic lipopeptide production by plant-associated *Pseudomonas* spp.: diversity, activity, biosynthesis, and regulation. *Molecular Plant-Microbe Interactions* 19: 699–710.

Roongsawang, N., Thaniyavarn, J., Thaniyavarn, S., et al. 2002. Isolation and characterization of a halotolerant *Bacillus subtilis* BBK-1 which produces three kinds of lipopeptides: bacillomycin L, plipastatin, and surfactin. *Extremophiles* 6: 499–506.

Roongsawang, N., Washio, K. and Morikawa, M. 2011. Diversity of nonribosomal peptide synthetases involved in the biosynthesis of lipopeptide biosurfactants. *International Journal of Molecular Sciences* 12: 141–172.

Sachdev, D. P. and Cameotra, S. S. 2013. Biosurfactants in agriculture. *Applied Microbiology and Biotechnology* 97: 1005–1016.

Šali, A. and Blundell, T. L. 1993. Comparative protein modelling by satisfaction of spatial restraints. *Journal of Molecular Biology* 234: 779–815.

Singh, S. S. 2006. Preclinical pharmacokinetics: an approach towards safer and efficacious drugs. *Current Drug Metabolism* 7: 165–182.

Singla, R. K., Dubey, H. D. and Dubey, A. K. 2014. Therapeutic spectrum of bacterial metabolites. *Indo Global Journal of Pharmaceutical Sciences* 2: 52–64.

Sivapathasekaran, C., Das, P., Mukherjee, S., et al. 2010. Marine bacterium derived lipopeptides: characterization and cytotoxic activity against cancer cell lines. *International Journal of Peptide Research and Therapeutics* 16: 215–222.

Tetko, I. V., Bruneau, P., Mewes, H.-W., Rohrer, D. C. and Poda, G. I. 2006. Can we estimate the accuracy of ADME–Tox predictions? *Drug Discovery Today* 11: 700–707.

Thaniyavarn, J., Roongsawang, N., Kameyama, T., et al. 2003. Production and characterization of biosurfactants from *Bacillus licheniformis* F2. 2. *Bioscience, Biotechnology, and Biochemistry* 67: 1239–1244.

Tropsha, A. 2010. Best practices for QSAR model development, validation, and exploitation. *Molecular Informatics* 29: 476–488.

Tropsha, A., Gramatica, P. and Gombar, V. K. 2003. The importance of being earnest: validation is the absolute essential for successful application and interpretation of QSPR models. *QSAR &Combinatorial Science* 22: 69–77.

Wang, L. and Xie, X.-Q. 2014. Computational target fishing: what should chemogenomics researchers expect for the future of *in silico* drug design and discovery? *Future Medicinal Chemistry* 6: 247–249.

Wang, T., Liang, Y., Wu, M., et al. 2015. Natural products from *Bacillus subtilis* with antimicrobial properties. *Chinese Journal of Chemical Engineering* 23: 744–754.

Wei, B. Q., Weaver, L. H., Ferrari, A. M., Matthews, B. W. and Shoichet, B. K. 2004. Testing a flexible-receptor docking algorithm in a model binding site. *Journal of Molecular Biology* 337: 1161–1182.

Wermuth, C., Ganellin, C., Lindberg, P. and Mitscher, L. 1998. Glossary of terms used in medicinal chemistry (IUPAC Recommendations 1998). *Pure and Applied Chemistry* 70: 1129–1143.

Xia, W., Du, Z., Cui, Q., et al. 2014. Biosurfactant produced by novel *Pseudomonas* sp. WJ6 with biodegradation of n-alkanes and polycyclic aromatic hydrocarbons. *Journal of Hazardous Materials* 276: 489–498.

Yoon, S.-H., Kim, J.-B., Lim, Y.-H., et al. 2005. Isolation and characterization of three kinds of lipopeptides produced by *Bacillus subtilis* JKK238 from Jeot-Kal of Korean traditional fermented fishes. *Microbiology and Biotechnology Letters* 33: 295–301.

Zhang, Y. 2008. Progress and challenges in protein structure prediction. *Current Opinion in Structural Biology* 18: 342–348.

Zouari, R., Hamden, K., El Feki, A., et al. 2016. Protective and curative effects of *Bacillus subtilis* SPB1 biosurfactant on high-fat-high-fructose diet induced hyperlipidemia, hypertriglyceridemia and deterioration of liver function in rats. *Biomedicine & Pharmacotherapy* 84: 323–329.

CHAPTER 2

Pore-Forming Antibacterial Lipopeptides

2.1 INTRODUCTION

Lipopeptides are microbial surface-active compounds produced by bacteria, yeast and fungi. The characteristic properties of lipopeptides are due to their high structural diversity. They possess the capability of decreasing interfacial and surface tension and are endowed with a diversified biological activity which makes them suitable as antimicrobial and antiviral drugs and insecticides (Muthusamy et al. 2008). The potential of pore formation and subsequent destabilization of the biological membrane permits their use as a potent antimicrobial, antitumor and hemolytic agent and promotes their use in the pharmaceutical, biomedical and agriculture fields. Their characteristic features are due to being an amphiphilic compound with hydrophobic (saturated or unsaturated fatty acid) and hydrophilic (amino acids or peptides, di- or polysaccharide, cations or anions) moieties. This feature corresponds to an isoform group in which there are variations in the composition of amino acids in peptide moiety, the length of the side chain of fatty acid and the bond between both the parts. The amphiphilic nature exerts a detergent-like property in lipopeptides, which is responsible for increasing the surface tension resulting in the process of pore formation in the bacterial cell wall and ultimately leading to cell death. Low critical micelle concentration (CMC) characterizes the lipopeptide surfactants. Critical micelle concentration is the minimum concentration of a detergent above which the self-assembly of monomers leads to non-covalent aggregates called micelles. This causes a sudden decrease in surface tension (McBain 1913). Similar phenomena describe the mechanism of action of lipopeptides used as an antibacterial drug. Lipopeptides with pore-forming properties can be categorized as calcium-dependent antibiotics (CDAs). CDAs are known to act upon the bacterial membrane wherein the formation of an oligomer of the drug molecule is supposed to induce perturbation in the membrane (Gandhimathi et al. 2009, Meca et al. 2011, Sharma et al. 2014). Calcium-dependent antibiotics described in this chapter such as friulimicin B and tsushimycin exert the bactericidal property through inhibition of biosynthesis of the bacterial cell wall. The inhibition occurs via formation of a complex with an essential precursor of the bacterial cell wall, i.e. undecaprenyl phosphate (C55-P) (McBain 1913, Peláez et al. 2011). Laspartomycin C also shares certain structural similarities with friulimicin B and tsushimycin. All of them contain amide-linked macrocycles, D-pipecolic acid and proline (Pro) residues at positions 3 and 11. Antibacterial activity of pore-forming antibiotics increases with the increased concentration of calcium because of their calcium dependency to exert the bactericidal property (Schneider et al. 2009).

2.2 FRIULIMICIN B

Friulimicin B is a lipopeptide-based antibiotic produced by *Actinoplanes friuliensis*. It is active against multiresistant Gram-positive bacteria such as strains of *Staphylococcus* and *Enterococcus*. The structure of friulimicin B consists of ten amino acids forming a ring structure and one exocyclic amino acid attached to an acyl residue (see Figure 2.1). The antibiotic targets the bacterial cell wall synthesis, which is possibly inhibited by an antibiotic-mediated complexation of the carrier bactoprenylphosphate. The structure of the antibiotic explains that an identical macrocyclic peptide is found as a central element in all lipopeptides. It is N-terminally attached through diaminobutyric acid (DAB) either to an aspartic acid or acylated asparagine residues (Vertesy et al. 2000). Apart from proteinogenic amino acids, friulimicin's peptide core also consists of some unusual amino acids such as pipecolinic acid, DAB and methylaspartic acid (L-threo--methylaspartic acid) (Vertesy et al. 2000). Reversible rearrangement of L-glutamate produces friulimicin. This is catalyzed by the enzyme glutamate mutase,

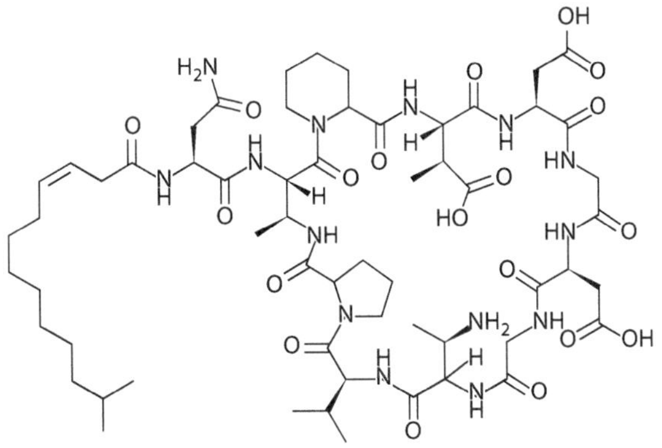

FIGURE 2.1 Chemical structure of antibiotic friulimicin B from *A. friuliensis.*

an adenosylcobalamin (coenzyme B12)-dependent enzyme (Heinzelmann et al. 2003).

2.2.1 Activity of Friulimicin B in Bacterial Cell

Friulimicin is an amphiphilic, water-soluble molecule with an overall negative charge. It requires calcium ions for its activity, which also enhances its amphiphilicity. This physiochemical property of friulimicin resembles that of daptomycin. Both of them kill Gram-positive bacteria by forming pores in the bacterial cytoplasmic membrane. Apart from this similarity with daptomycin, friulimicin has a distinct mechanism of action. Bactoprenol phosphate is a lipid carrier for the biosynthesis of cell wall teichoic acid and helps in the transport of polysaccharide across the cytoplasmic membrane. Thus, it is required for cell wall synthesis in Gram-positive bacteria. Friulimicin forms a calcium-dependent complex with bactoprenol phosphate carrier, thereby interrupting the cell wall precursor cycle and blocking the cell envelope synthesis of Gram-positive bacteria (Schneider et al. 2009).

Undecaprenylphosphate (C55-P) has a role in capsule formation and serves as a carrier in the biosynthesis of teichoic acid. This explains the fact that friulimicin acts on blocking multiple pathways essential for the functioning of the Gram-positive cell envelope. It is found that without affecting the membrane integrity, friulimicin forms a complex with bactoprenol phosphate. Apart from the function of cell wall biosynthesis, C55-P also plays a role of lipid carrier for the biosynthesis of wall teichoic acid and polysaccharide transport across the cytoplasmic membrane. So, in Gram-positive pathogens, precursor cycling is interrupted through the abduction of C55-P carrier and as a result, functional cell envelope synthesis is blocked (Schneider et al. 2009).

2.2.2 Ligands of Friulimicin B

Ligands are the small molecules which are considered as potential drug targets because of their specific affinity toward a particular drug molecule. These small molecules can be present on the cell wall of bacteria. As a part of the bacterial cell wall these molecules might play a vital role in the process of biosynthesis. There is evidence that such drugs interact with their unique ligands, i.e. small molecules such as amino acids (e.g. tryptophan, a unique ligand known as 4FO) as part of a protein molecule present on the cell wall, and molecules like DAB, as part of the cell wall biosynthesis process. Such molecules can interact with the exposed drug molecule in order to treat the infection of the respective microorganism by their inhibitory action. The infection can be treated if the drug and small molecule present on the cell wall have some interaction. As discussed in the mechanism of action of friulimicin B, DAB has a role in the cell wall biosynthesis process of bacteria. DAB and 4FO are taken as unique ligands for lipopeptides-based drug, friulimicin B, to demonstrate the drug–ligand interaction and inhibitory effect of the drug. Ligand data is retrieved from the protein data bank (PDB) available on the RCSB website (www.rcsb.org/). Two-dimensional structures of ligands have been drawn using a software-based tool named two-dimensional (2D) sketcher available on Maestro suite of Schrödinger software. A list of ligands used for molecular docking with their IUPAC name and 2D structures are shown in Table 2.1.

TABLE 2.1   Ligands of Friulimicin B with 2D Structures

Ligand	IUPAC Name & Molecular Formula	2D Structure
DAB	2,4-Diaminobutyric acid $C_4 H_{10} N_2 O_2$	
4FO	D-Tryptophan $C_{11} H_{12} N_2 O_2$	

2.2.3 Docking Studies for Friulimicin

Molecular docking is an essential part of drug design. Molecular docking is performed to show the interaction between the drug and the ligand molecule. The process of docking consists of several steps in which the first step is the preparation of protein molecule, which includes the preparation of all the possible poses of a three-dimensional (3D) protein molecule. Site generation is performed to originate suitable sites on the protein molecule for creating major and minor grooves. This generates a 3D space for the ligand molecule to interact with the electrons of surrounding amino acids. This interaction between the ligand and the drug molecule is very important and the results are generated by the software in the form of docking score and glide energy which gives the information of binding affinity of participating molecules. Molecular docking study exhibits the affinity of drug and ligand in the form of hydrogen bond interaction. The 3D structure of friulimicin B used for docking is generated using homology modeling because the 3D structure of friulimicin B is not available in PDB. Docking results are generated using the Maestro suite of Schrödinger software. Glide is an application of Maestro suite, which is used to perform docking. The results of a docking study for lipopeptide-based drug friulimicin B are depicted in Table 2.2, which shows that the docking score of friulimicin B is –3.405 and –3.375 for DAB and 4FO, respectively. The glide energy of docking of friulimicin B and DAB is –20.475 kcal/mol (see Figure 2.2). Interaction of friulimicin B and 4FO gave the glide energy score of –20.559 kcal/mol (see Figure 2.3). DAB interacted with 2 hydrogen bonds with friulimicin B (see Figure 2.2 B) and 4FO interacted with 4 hydrogen bonds (see Figure 2.3 B). As per the protocol, the lowest docking score is considered as the best result. Here, according to the data generated, DAB has the best docking with friulimicin B, because it has the lowest docking score which is independent of hydrogen bond interaction and glide energy. The ligand interaction diagram (LID) is generated by the software as shown in Figures 2.2 A and 2.3 A and hydrogen bond

TABLE 2.2   Docking Results of Friulimicin B with its Ligands

Ligand	DAB	4FO
H Bond	2	4
Docking Score	–3.405	–3.375
Glide Energy (kcal/mol)	–20.475	–20.559

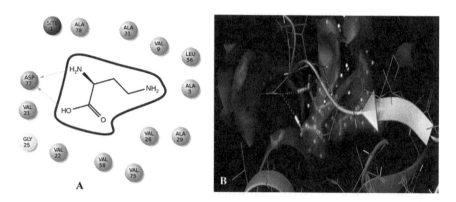

FIGURE 2.2   Ligand interaction diagram (A) and H-bond interaction diagram of friulimicin and DAB (B).

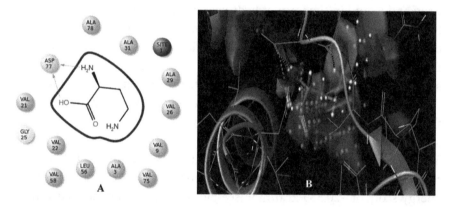

FIGURE 2.3   Ligand interaction diagram (A) and H-bond interaction diagram of friulimicin and 4FO (B).

interaction is shown as dotted lines in Figures 2.2 B and 2.3 B for ligand molecule DAB and 4FO, respectively.

2.2.4 ADMET Study for Friulimicin

ADMET study is the assessment of pharmacokinetics of a drug which stands for Absorption, Distribution, Metabolism, Excretion and Toxicity. The prediction of the fate of a drug and the effects caused by a drug inside the body, such as how much drug is absorbed if administered orally and how much is absorbed in the gastrointestinal tract, is an indispensable part of drug discovery. In a similar way, if the absorption is poor, its distribution and metabolism would be affected, which can lead to causing neurotoxicity and nephrotoxicity. Thus, ADMET study is the most essential part of computational drug design. ADMET study can be performed using various web-based tools. Here, ADMET study is performed using a web-based tool named admetSAR. This tool predicts the ADMET properties of the drug with the following descriptors, that are given in Table 2.3.

TABLE 2.3 ADMET Results of Friulimicin B

ADMET Predicted Profile – Classification		
Model	Result	Probability
Absorption		
Blood–Brain Barrier	BBB-	0.9970
Human Intestinal Absorption	HIA-	0.7560
Caco-2 Permeability	Caco2-	0.8189
P-glycoprotein Substrate	Substrate	0.7731
P-glycoprotein Inhibitor	Non-inhibitor	0.6863
	Non-inhibitor	0.8718
Renal Organic Cation Transporter	Non-inhibitor	0.9338
Distribution		
Subcellular Localization	Lysosome	0.4858
Metabolism		
CYP450 2C9 Substrate	Non-substrate	0.8407
CYP450 2D6 Substrate	Non-substrate	0.8086
CYP450 3A4 Substrate	Substrate	0.5961
CYP450 1A2 Inhibitor	Non-inhibitor	0.9598
CYP450 2C9 Inhibitor	Non-inhibitor	0.9075
CYP450 2D6 Inhibitor	Non-inhibitor	0.9481
CYP450 2C19 Inhibitor	Non-inhibitor	0.9328
CYP450 3A4 Inhibitor	Non-inhibitor	0.9887
CYP Inhibitory Promiscuity	Low CYP Inhibitory Promiscuity	0.9902
Excretion		
Toxicity		
Human Ether-a-go-go-Related Gene Inhibition	Weak Inhibitor	0.9770
	Non-inhibitor	0.8863
AMES Toxicity	Non AMES toxic	0.8265
Carcinogens	Non-carcinogens	0.9228
Fish Toxicity	High FHMT	0.9516
Tetrahymena pyriformis Toxicity	High TPT	0.9606
Honey Bee Toxicity	Low HBT	0.7354
Biodegradation	Not ready biodegradable	0.9831
Acute Oral Toxicity	III	0.5746
Carcinogenicity (Three-class)	Non-required	0.6290
		(Continued)

TABLE 2.3 (CONTINUED)   ADMET Results of Friulimicin B

Model	ADMET Predicted Profile – Regression	
	Value	Unit
Absorption		
Aqueous Solubility	−2.5422	LogS
Caco-2 Permeability	−0.4742	LogPapp, cm/s
Distribution		
Metabolism		
Excretion		
Toxicity		
Rat Acute Toxicity	2.9688	LD50, mol/kg
Fish Toxicity	1.7888	pLC50, mg/L
Tetrahymena pyriformis Toxicity	0.3211	pIGC50, ug/L

2.2.5 Pharmacophore Study for Friulimicin

Pharmacophore study is used to perform structural alignment and defines molecular similarity. A pharmacophore model depicts the spatial arrangement of vital features of an interaction. Each pharmacophore query feature includes both a type and a radius. Pharmacophore study is performed using a web-based tool named "Pharmit." The pharmacophore property derives functional groups of the underlying chemistry and in the perspective of software "Pharmit," it is defined as a hydrogen acceptor, hydrogen donor, aromatic, hydrophobic, positive ion or negative ion. Pharmacophore features are selected to represent a particular ligand and their geometrical orientation. The radius of a pharmacophore query feature determines how closely a molecule in the database should match the configuration of the query. Pharmacophore properties are a set of features provided in the "Pharmacophore" menu in sidebar. They can be toggled "on" or "off" to incorporate them or eliminate them from the query search. Pharmacophore models generated for lipopeptides friulimicin with its respective ligand are shown in Figure 2.4.

2.3 TRIDECAPTIN A

Tridecaptin A, also referred as TriA1, is a non-ribosomal lipopeptide produced by strains of *Bacillus polymyxa*. It possesses selective antimicrobial activity against a wide category of Gram-negative bacteria. Tridecaptin is a family of antibacterial drugs from a non-ribosomal lipopeptide group. These are generally produced by *Paenibacillus* and *Bacillus* species

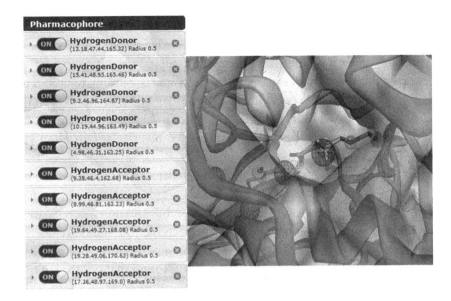

FIGURE 2.4   Pharmacophore model for homologous structure of friulimicin B with ligand PNS generated by Pharmit.

(Shoji et al. 1978, Lohans et al. 2014, Cochrane et al. 2015). The acylated tridecaptin possesses selective and strong antimicrobial activity against a category of Gram-negative bacteria which includes multidrug-resistant strains of *Escherichia coli*, *Acinetobacter baumannii* and *Klebsiella pneumonia* (Cochrane et al. 2014).

2.3.1 Structure of Tridecaptin A

The structure of tridecaptin A was determined by gas chromatography and mass spectrometry and was found to have the following amino acids: 2,4-DAB (2D, 1L), Glu (1L), Ser (1D,1L), Gly (1), Val (1D, 1L), Ala (1L), alle (1D), Phe (1L) and Trp (1D) with 8-hydroxy anteisononanoic acid as the fatty acid (Shoji et al. 1976). The NMR structure of tridecaptin A1 in solution was found to contain dodecylphosphocholine micelles with lipids II. To depict a structural model of the TriA1-lipids II complex, molecular modeling was used. For clarification of the structure of tridecaptin A, further evidences such as cleavage reaction with N-bromosuccinimide, partial acid hydrolysis, sequential analysis by Edman degradation and certain other examinations were also performed. Tridecaptin A, an acyl peptide antibiotic, yields the previously listed constituent amino acids and fatty acid on acid hydrolysis. These degradation studies elucidate the structure of Tridecaptin A (Kato et al. 1978).

2.3.2 Mode of Action of Tridecaptin A

The bactericidal property of Tridecaptin is exerted by its binding to the precursor lipid bacterial cell wall resulting in disrupted proton motive force. According to the biophysical and biochemical assays, binding to lipid II of Gram-negative variant plays a vital role in the disruption of the membrane and in the dispersion of proton gradient. The studies explain that Gram-negative bacteria are killed by TriA1 through a mechanism of action via a lipid-II-binding motif. The outer membrane of Gram-negative bacteria protects the bacteria from the surrounding environment. The outer membrane takes part in reducing diffusion of various antibiotics into the periplasmic space of the cell. Tridecaptin must cross the outer barrier first to attack the membrane of Gram-negative bacteria. The structure–activity relationship of TriA1 shows that it has membrane-targeting activity, similar to other pore-forming lipopeptides. It is found that removal of the lipid tail located at N-terminal destroys the antimicrobial activity. This shows the importance of the lipid tail for the antimicrobial activity of tridecaptin. Even though the chiral lipid tail replaced with an octanoyl chain produces a Oct-TriA1 molecule, full activity of tridecaptin is still retained (Cochrane et al. 2014). This helped in identifying the precise mode of action of Tridecaptin to kill Gram-negative bacteria. As per the previous studies, it was assumed that tridecaptin, like polymyxin, also crosses the outer membrane and traverses inside the microorganism by interaction with lipopolysaccharide (LPS) (Velkov et al. 2009) as it contains some cationic residues like 2, 4-DAB. This LPS binding is the mechanistic basis for the disruption of bacterial outer membrane.

2.3.3 Ligands of Tridecaptin A

Unique ligands are the small molecules with selective affinity for a particular drug. Mostly, these ligands are some derivatives of amino acids. Such amino acids are also part of proteins found on the cell wall of bacteria. If bacterial cell wall is exposed with a drug molecule, the ligands present on cell wall interact with the drug because of the specific affinity of such amino acids toward the drug molecule. This interaction leads to the disruption of bacterial cell wall by which the drug molecule ultimately exhibits its mode of action. The ligands used for docking study are retrieved from Protein Data Bank (PDB) available on RCSB (www.rcsb.org/). The experimental data of the 3D crystal structure of tridecaptin is available in PDB with PDB ID 2N5Y. Unique ligands for 2N5Y are taken which are listed in Table 2.4.

TABLE 2.4   Unique Ligands with 2D Structure for Tridecaptin A

Ligand	IUPAC Name & Molecular Formula	2D Structure
28J	D-alloisoleucine $C_6 H_{13} N O_2$	
4FO	(2R)-2,4-diaminobutanoic acid $C_4 H_{10} N_2 O_2$	
4N3	N-octanoyl-D-valine $C_{13} H_{25} N O_3$	
DAB	2,4-Diaminobutyric acid $C_4 H_{10} N_2 O_2$	
DSN	D-Serine $C_3 H_7 N O_3$	
DTR	D-Tryptophan $C_{11} H_{12} N_2 O_2$	

2.3.4 Molecular Docking Studies of Tridecaptin

The Glide application of Maestro suit was used to perform ligand docking. Results are described in terms of the number of hydrogen bonds
formed between the drug and ligand. The docking score and glide energy
are noted to analyze the interaction. The docking score is the value that
needs to be considered for the best interaction as the lowest docking score
is considered as the best docking result. Docking results are described in
Table 2.5 with the values of number of hydrogen bonds formed, docking
score and glide energy generated by the formation of a complex between
given ligands and drug molecule during docking process. The ligand

28J interacted with tridecaptin via one hydrogen bond with the docking score and glide energy of −2.459 and −17.956 kcal/mol, respectively (see Figure 2.5). 4FO had a docking score of −3.132 and a glide energy of −20.154 kcal/mol. The interaction resulted in the formation of three hydrogen bonds (see Figure 2.6). DAB formed three hydrogen bonds with tridecaptin and resulted in a docking score of −2.921 and glide energy of −17.602 kcal/mol (see Figure 2.7). Ligand DSN docked with tridecaptin with a score and glide energy of −2.112 and −19.266 kcal/mol, respectively with two hydrogen bonds (see Figure 2.8). Tridecaptin has the best docking score with the ligand DTR (D-Tryptophan), i.e. −3.953, which is the lowest docking score and glide energy is −25.131 kcal/mol. The interaction between tridecaptin and DTR (formation of three hydrogen bonds) can be visualized as three dotted lines in Figure 2.9.

TABLE 2.5   Molecular Docking Results of Tridecaptin A

Ligand	28J	4FO	DAB	DSN	DTR
H Bond	1	3	3	2	3
Docking Score	−2.459	−3.132	−2.921	−2.112	−3.953
Glide Energy (kcal/mol)	−17.956	−20.154	−17.602	−19.266	−25.131

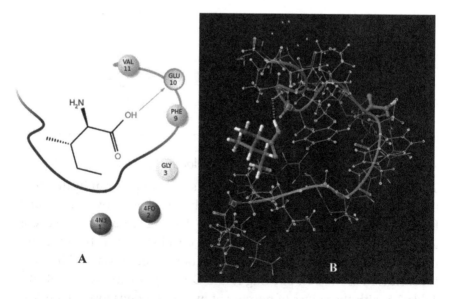

FIGURE 2.5   Ligand interaction diagram (A) and hydrogen bond interaction of tridecaptin and 28J (B).

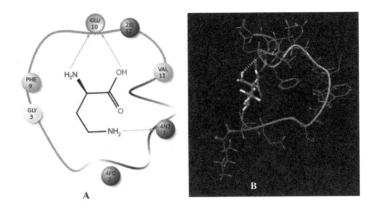

FIGURE 2.6   Ligand interaction diagram (A) and hydrogen bond interaction of tridecaptin and 4FO (B).

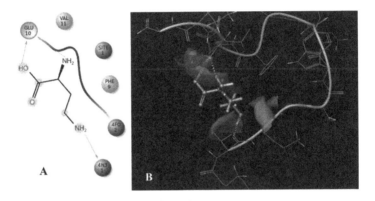

FIGURE 2.7   Ligand interaction diagram (A) and hydrogen bond interaction of tridecaptin and DAB (B).

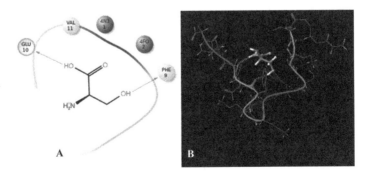

FIGURE 2.8   Ligand interaction diagram (A) and hydrogen bond interaction between tridecaptin and DSN (B).

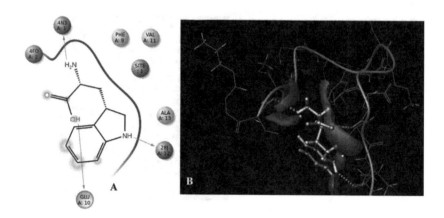

FIGURE 2.9   Ligand interaction diagram (A) and hydrogen bond interaction between tridecaptin and DTR (B).

2.3.5 ADMET Properties

ADMET is an abbreviation of pharmacological activities that contribute to the disposition of a drug in an organism. The absorption of a drug, its distribution and the metabolism of a drug inside the body play a vital role in the process of drug discovery. ADMET is performed using software-based tools in computer-aided drug design to determine the fate of the drug. Certain tools are available online which calculate the results on the basis of predefined parameters. ADMET study for tridecaptin is performed using a web-based tool named admetSAR which is available on the website lmmd.ecust.edu.cn:8000/. The software-based study predicts the descriptors which are mentioned in Table 2.6.

TABLE 2.6   ADMET Results for Tridecaptin A

ADMET Predicted Profile – Classification		
Model	**Result**	**Probability**
Absorption		
Blood–Brain Barrier	BBB-	0.9503
Human Intestinal Absorption	HIA+	0.9639
Caco-2 Permeability	Caco2-	0.8343
P-glycoprotein Substrate	Substrate	0.8438
P-glycoprotein Inhibitor	Non-inhibitor	0.8495
	Non-inhibitor	0.7967
Renal Organic Cation Transporter	Non-inhibitor	0.8743
Distribution		
Subcellular Localization	Mitochondria	0.4623
Metabolism		
CYP450 2C9 Substrate	Non-substrate	0.8623
CYP450 2D6 Substrate	Non-substrate	0.7557
CYP450 3A4 Substrate	Substrate	0.5383
CYP450 1A2 Inhibitor	Non-inhibitor	0.8869
CYP450 2C9 Inhibitor	Non-inhibitor	0.8374
CYP450 2D6 Inhibitor	Non-inhibitor	0.8403
CYP450 2C19 Inhibitor	Non-inhibitor	0.7999
CYP450 3A4 Inhibitor	Non-inhibitor	0.5983
CYP Inhibitory Promiscuity	Low CYP Inhibitory Promiscuity	0.6724
Excretion		
Toxicity		
Human Ether-a-go-go-Related Gene Inhibition	Weak Inhibitor	0.9904
	Non-inhibitor	0.5592
AMES Toxicity	Non AMES Toxic	0.8368
Carcinogens	Non-carcinogens	0.8734
Fish Toxicity	High FHMT	0.9800
Tetrahymena pyriformis Toxicity	High TPT	0.9963
Honey Bee Toxicity	Low HBT	0.7160
Biodegradation	Not Ready biodegradable	0.9787
Acute Oral Toxicity	III	0.6128
Carcinogenicity (Three-class)	Non-required	0.6390
ADMET Predicted Profile – Regression		
Model	**Value**	**Unit**
Absorption		
Aqueous Solubility	−3.6526	LogS
Caco-2 Permeability	−0.2460	LogPapp, cm/s
		(*Continued*)

TABLE 2.6 (CONTINUED)   ADMET Results for Tridecaptin A

	Distribution Metabolism Excretion Toxicity	
Rat Acute Toxicity	3.0062	LD50, mol/kg
Fish Toxicity	1.4323	pLC50, mg/L
Tetrahymena pyriformis Toxicity	0.5092	pIGC50, ug/L

2.3.6 Concept of Pharmacophore for Tridecaptin A

Pharmacophore modeling is conducted using "Pharmit," which is an online tool for generating a pharmacophore model of a query protein. Here, lipopeptides are taken to see the pharmacophore properties to design lipopeptides-based drugs. The standard bond length and radius around atoms in the interaction are shown in figures. Pharmacophore modeling for tridecaptin is depicted in Figures 2.10–2.12.

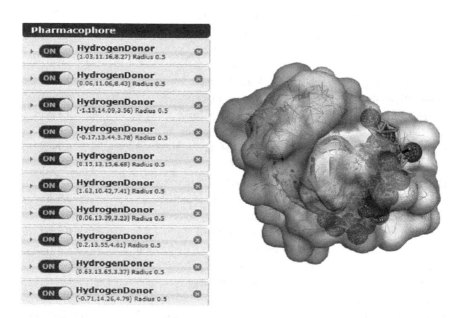

FIGURE 2.10   Pharmacophore model of tridecaptin with ligand 28J.

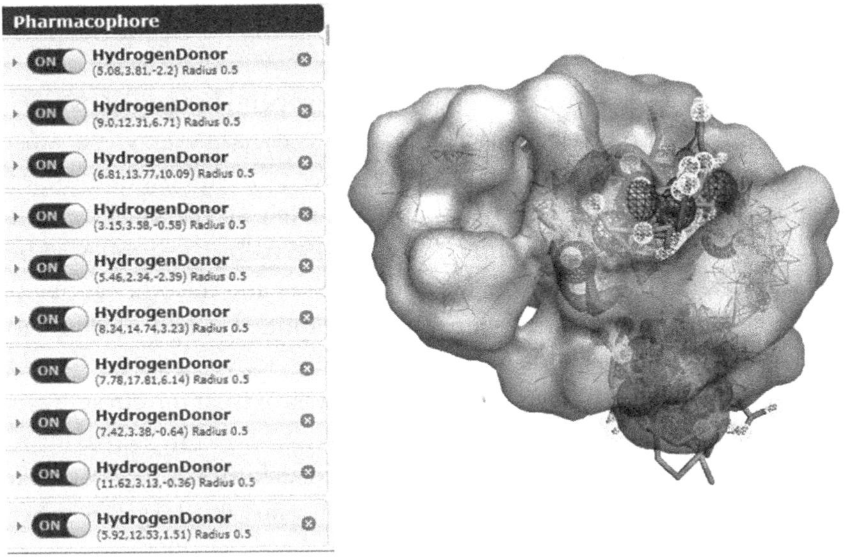

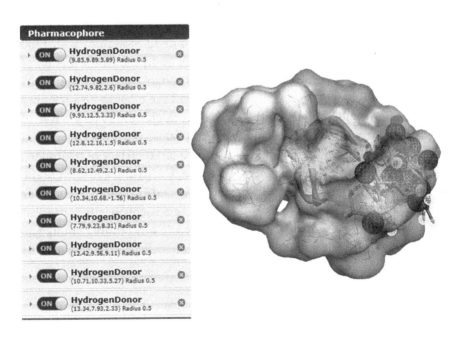

FIGURE 2.11   Pharmacophore model of tridecaptin with 4FO.

FIGURE 2.12   Pharmacophore study of tridecaptin and ligand DAB.

2.4 TSUSHIMYCIN

2.4.1 Introduction of Tsushimycin

Tsushimycin is a derivative of amphomycin. Lipopeptide antibiotics comprise a family of antibacterial drugs which shows dynamic activity against various multidrug-resistant bacteria. In the series of lipopeptides-based antibiotics, amphomycin, discovered approximately 50 years ago, was the first member of the series (Heinemann et al. 1953). This was followed by the identification of tsushimycin (Shoji et al. 1968) and the other antibiotics which belong to the same group. This lipopeptide antibiotic belongs to the class of calcium-dependent antibiotics and is active against a class of Gram-positive bacteria. Crystal structure analysis of tsushimycin suggests that the antibiotic is likely to be a dimer in its biologically active form. Dimerization depends upon the presence of calcium ion (Ca^{2+}) which accommodates the structure to be suitable for a possible target of either neutral or acidic nature (Schneider 2002, Bunkóczi et al. 2005). The crystal structure of tsushimycin was determined at a resolution of 1.0 Å. The structure contains a peptide framework consisting of a cyclopeptide ring of ten members and an exocyclic amino acid, the $-NH_2$ group of which is acylated by a fatty acid residue (Bunkóczi et al. 2005). This is one of the largest structures solved by ab initio direct methods. The structure contains an asymmetric unit which is composed of 12 molecules with 1300 independent atoms. The backbone is found to be in the conformation of a saddle-like structure which is stabilized through a calcium ion (Ca^{2+}) bound inside the peptide ring. This categorizes the drug to be a member of calcium-dependent antibiotics. Additional calcium ions play a role of linking the drug molecule to dimers which enclose the empty space and resemble a binding cleft. The dimers acquire a large surface of hydrophobic nature which has the capability to interact with the cell membrane of bacteria. The most studied drug, daptomycin, also exhibits the similar conformation because the constituent amino acid sequence is conserved at positions which are involved in calcium ion (Ca^{2+}) binding (Bunkóczi et al. 2005). This series of antibiotics principally differ in the structure of their fatty acid substituent.

2.4.2 Physiological Effect of Tsushimycin

The mechanism of action for tsushimycin is not defined clearly. But as per the early studies, it is indicated that tsushimycin is a specific inhibitor of biosynthesis of bacterial cell wall. The level at which tsushimycin acts via inhibition of cell wall synthesis is MraY (Phosphoacetylmuramoyl-pentapeptide transferase), which acts as a catalyzer for transferring UDP-MurNac-pentapeptide to the undecaprenyl carrier (Tanaka et al. 1982). The formation of nucleotide-linked sugar-peptide precursor is the stage for tsushimycin to act as an inhibitor of peptidoglycan biosynthesis (Allen et al. 1987). The level of calcium ion (Ca^{2+}) in the medium has proved to be important in testing against the isolates which are less susceptible. In trials on calcium (Ca^{2+}) supplemented media, the value of MIC (minimum inhibitory concentration) was twofold to fourfold lower as compared to a medium with physiological calcium levels (Ca^{2+}). The specific antibacterial activity of lipopeptides against microorganisms makes them suitable for reinforcing glycopeptides as a last line therapy against lethal bacterial infections (Bunkóczi et al. 2005).

2.4.3 Identification of Ligands of Tsushimycin

Identification of ligands is an important step in recognizing the specific site of action of a drug molecule to distinguish antibacterial effect of a drug. Unique ligands bind specifically with their corresponding drugs. The data of unique ligands is provided by the PDB available on RCSB (www.rcsb.org/). The 3D crystal structure of tsushimycin is available in the PDB database. The PDB ID of Tsushimycin 1W3M was used for molecular docking. Ligand identification and molecular docking are the essential steps in the process of drug discovery. 2D structures of ligands were drawn using a software tool named 2D sketcher available on the Maestro suite of Schrödinger software. The advantage of this software is that the structures drawn using 2D sketcher are converted into 3D conformation. The list of unique ligands for tsushimycin is given in Table 2.7.

TABLE 2.7   Ligands of Tsushimycin with Their IUPAC Names and 2D Structures

Ligand	IUPAC Name & Molecular Formula	2D Structure
LNG	δ-3isotetradecenoic acid $C_{14} H_{26} O_2$	
EOH	Ethanol $C_2 H_6 O$	HO⎯⎯CH₃
CA	Calcium ion Ca^{2+}	Ca^{2+}
CL	Chloride ion Cl^-	Cl^-

2.4.4 Molecular Docking Simulations

The molecular docking method is used to predict the favorable orientation of participating molecules to bind to each other for the formation of a stable complex (Lengauer and Rarey 1996). If the preferred orientation is known, this can predict the binding affinity or strength of the complex between two molecules. Molecular docking is an essential part of computer-aided drug design and structural molecular biology. The objective of drug–ligand docking is to predict the preferred modes of binding of a drug molecule with known 3D conformations with the ligand. The aim of molecular docking is to depict how a target is inhibited by the drug molecule. This is an indispensable part of lead optimization (Schneider 2002). Docking is performed by using the Maestro suite of Schrödinger software that facilitates an application named Glide which is used to perform ligand docking. To perform docking the 3D structure of tsushimycin was retrieved from the PDB database with the PDB ID 1W3M in .pdb format. The results of docking are given in Table 2.8. The docking score and glide energy of tsushimycin with LNG is −3.323 and −25.669 kcal/mol, respectively. The number of hydrogen bonds formed during interaction is one (see Figure 2.13). Schrödinger software also facilitates generation of the ligand interaction diagram (LID). A LID for each docking that is used to visualize the interaction of electrons of a ligand with participating amino acids on an available site for docking. Hydrogen bond interaction can be seen as dotted lines formed during the interaction. The LID and hydrogen bond interaction diagram between tsushimycin and LNG is shown in Figure 2.13 (A) and (B), respectively. Ethanol, calcium ion and chloride ion were not considered for docking as they are very small molecules as compared to the drug molecule tsushimycin which consists of a 132 residue count and a 1125 atom count (Bunkóczi et al. 2005).

2.4.5 ADMET Properties of Tsushimycin

ADMET properties of a drug are studied to conduct the assessment of hazards related to the human and the environment (Cheng et al. 2012).

TABLE 2.8   Docking Result of Tsushimycin

Ligand	LNG
H Bond	1
Docking Score	−3.323
Glide Energy (kcal/mol)	−25.669

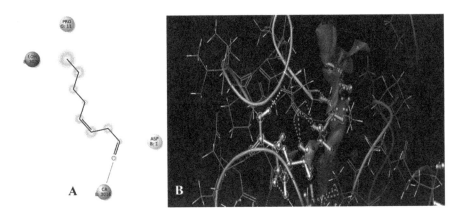

FIGURE 2.13 Ligand interaction diagram (A) and hydrogen bond interaction between tsushimycin and ligand LNG (B).

The disposition of drug inside the body and the after effects generated post-excretion in the environment such as toxicity and carcinogenecity depend upon various parameters which can be calculated using software-based analysis. ADMET properties of tsushimycin performed through the software admetSAR generated the results which need to be in the acceptable range. The data are shown in Table 2.9.

TABLE 2.9 ADMET Results for Tsushimycin

ADMET Predicted Profile – Classification		
Model	Result	Probability
Absorption		
Blood–Brain Barrier	BBB-	0.9970
Human Intestinal Absorption	HIA-	0.7560
Caco-2 Permeability	Caco2-	0.8189
P-glycoprotein Substrate	Substrate	0.7731
P-glycoprotein Inhibitor	Non-inhibitor	0.6863
	Non-inhibitor	0.8718
Renal Organic Cation Transporter	Non-inhibitor	0.9338
Distribution		
Subcellular Localization	Lysosome	0.4858
Metabolism		
CYP450 2C9 Substrate	Non-substrate	0.8407
CYP450 2D6 Substrate	Non-substrate	0.8086
CYP450 3A4 Substrate	Substrate	0.5961
CYP450 1A2 Inhibitor	Non-inhibitor	0.9598

(Continued)

TABLE 2.9 (CONTINUED)   ADMET Results for Tsushimycin

CYP450 2C9 Inhibitor	Non-inhibitor	0.9075
CYP450 2D6 Inhibitor	Non-inhibitor	0.9481
CYP450 2C19 Inhibitor	Non-inhibitor	0.9328
CYP450 3A4 Inhibitor	Non-inhibitor	0.9887
CYP Inhibitory Promiscuity	Low CYP Inhibitory Promiscuity	0.9902
	Excretion	
	Toxicity	
Human Ether-a-go-go-Related	Weak Inhibitor	0.9770
Gene Inhibition	Non-inhibitor	0.8863
AMES Toxicity	Non AMES toxic	0.8265
Carcinogens	Non-carcinogens	0.9228
Fish Toxicity	High FHMT	0.9516
Tetrahymena pyriformis Toxicity	High TPT	0.9606
Honey Bee Toxicity	Low HBT	0.7354
Biodegradation	Not Ready Biodegradable	0.9831
Acute Oral Toxicity	III	0.5746
Carcinogenicity (Three-class)	Non-required	0.6290

ADMET Predicted Profile – Regression

Model	Value	Unit
Absorption		
Aqueous Solubility	−2.5422	LogS
Caco-2 Permeability	−0.4742	LogPapp, cm/s
Distribution		
Metabolism		
Excretion		
Toxicity		
Rat Acute Toxicity	2.9688	LD50, mol/kg
Fish Toxicity	1.7888	pLC50, mg/L
Tetrahymena pyriformis Toxicity	0.3211	pIGC50, ug/L

2.4.6 Pharmacophore Studies of Tsushimycin with its Ligands

The crystal structure of tsushimycin is available as PDB entry 1W3N in PDB available on the RCSB website. For the purpose of drug design of lipopeptide *in silico* analysis is performed using various computational tools. Pharmacophore modeling is an indispensable step in computer-aided drug design. A pharmacophore model for tsushimycin depicting surface properties and the nature of interacting molecules is shown in Figure 2.14.

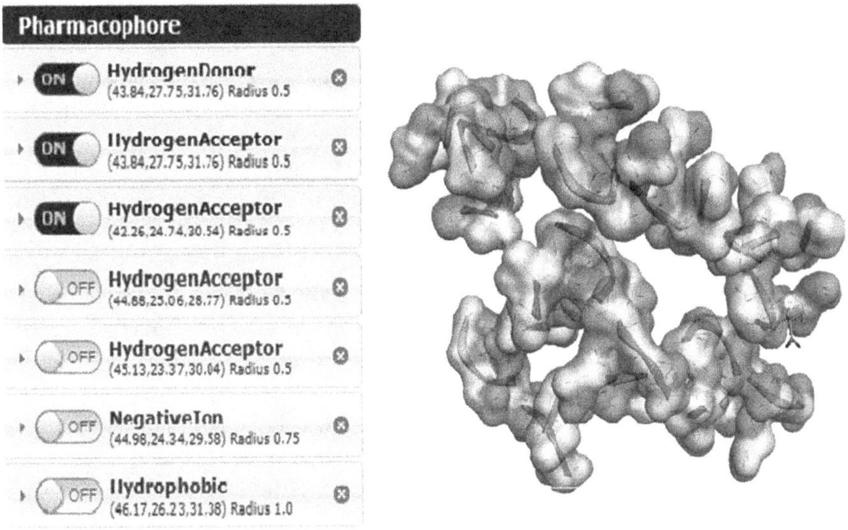

Pharmacophore

▸ **ON** ◯ **HydrogenDonor** ✖
(43.84,27.75,31.76) Radius 0.5

▸ **ON** ◯ **HydrogenAcceptor** ✖
(43.84,27.75,31.76) Radius 0.5

▸ **ON** ◯ **HydrogenAcceptor** ✖
(42.26,24.74,30.54) Radius 0.5

▸ ◯ OFF **HydrogenAcceptor** ✖
(44.68,25.06,28.77) Radius 0.5

▸ ◯ OFF **HydrogenAcceptor** ✖
(45.13,23.37,30.04) Radius 0.5

▸ ◯ OFF **NegativeIon** ✖
(44.98,24.34,29.58) Radius 0.75

▸ ◯ OFF **Hydrophobic** ✖
(46.17,26.23,31.38) Radius 1.0

FIGURE 2.14   Pharmacophore mapping of tsushimycin with ligand 2AS.

2.5 SUMMARY

In the current chapter, pore-forming lipopeptides were considered as potential drugs for drug-resistant microorganisms. These lipopeptides form pores in the bacterial cell membrane, thereby lysing and killing the cell. A few of the pore-forming lipopeptides are friulimicin, tridecaptin and tsushimycin. These lipopeptides were tested *in silico* for their efficacy as drugs. Among the two ligands of friulimicin (DAB and 4FO), DAB exhibits better docking which suggests that friulimicin is likely to show its pore-forming ability if DAB is present on the bacterial surface. Similarly, the presence of DTR will bind to tridecaptin A and aid in bacterial cell lysis. The final choice of the drug will depend upon the infective microorganism as friulimicin is active against Gram-positive bacteria whereas tridecaptin A targets Gram-negative bacterial species.

REFERENCES

Allen, N., Hobbs, J. and Alborn, W. 1987. Inhibition of peptidoglycan biosynthesis in gram-positive bacteria by LY146032. *Antimicrobial Agents and Chemotherapy* 31: 1093–1099.

Bunkóczi, G., Vértesy, L. and Sheldrick, G. M. 2005. Structure of the lipopeptide antibiotic tsushimycin. *Acta Crystallographica Section D: Biological Crystallography* 61: 1160–1164.

Cheng, F., Li, W., Zhou, Y., et al. 2012. *admetSAR: A Comprehensive Source and Free Tool for Assessment of Chemical ADMET Properties*, ACS Publications.

Cochrane, S. A., Lohans, C. T., Brandelli, J. R., et al. 2014. Synthesis and structure–activity relationship studies of N-terminal analogues of the antimicrobial peptide tridecaptin A1. *Journal of Medicinal Chemistry* 57: 1127–1131.

Cochrane, S. A., Lohans, C. T., van Belkum, M. J., Bels, M. A. and Vederas, J. C. 2015. Studies on tridecaptin B 1, a lipopeptide with activity against multidrug resistant Gram-negative bacteria. *Organic &Biomolecular Chemistry* 13: 6073–6081.

Gandhimathi, R., Kiran, G. S., Hema, T., et al. 2009. Production and characterization of lipopeptide biosurfactant by a sponge-associated marine actinomycetes *Nocardiopsis alba* MSA10. *Bioprocess and Biosystems Engineering* 32: 825–835.

Heinemann, B., Kaplan, M., Muir, R. and Hooper, I. 1953. Amphomycin, a new antibiotic. *Antibiotics &Chemotherapy (Northfield, Ill.)* 3: 1239.

Heinzelmann, E., Berger, S., Puk, O., et al. 2003. A glutamate mutase is involved in the biosynthesis of the lipopeptide antibiotic friulimicin in *Actinoplanes friuliensis*. *Antimicrobial Agents and Chemotherapy* 47: 447–457.

Kato, T., Hinoo, H. and Shoji, J. I. 1978. The structure of tridecaptin A. *The Journal of Antibiotics* 31: 652–661.

Lengauer, T. and Rarey, M. 1996. Computational methods for biomolecular docking. *Current Opinion in Structural Biology* 6: 402–406.

Lohans, C. T., van Belkum, M. J., Cochrane, S. A., et al. 2014. Biochemical, structural, and genetic characterization of tridecaptin A1, an antagonist of *Campylobacter jejuni*. *ChemBioChem* 15: 243–249.

McBain, J. 1913. Mobility of highly-charged micelles. *Transactions of the Faraday Society* 9: 99–101.

Meca, G., Sospedra, I., Valero, M. A., et al. 2011. Antibacterial activity of the enniatin B, produced by *Fusarium tricinctum* in liquid culture, and cytotoxic effects on Caco-2 cells. *Toxicology Mechanisms and Methods* 21: 503–512.

Muthusamy, K., Gopalakrishnan, S., Ravi, T. K. and Sivachidambaram, P. 2008. Biosurfactants: properties, commercial production and application. *Current Science* 94: 736–747.

Peláez, F., Collado, J., Platas, G., et al. 2011. Phylogeny and intercontinental distribution of the pneumocandin-producing anamorphic fungus *Glarea lozoyensis*. *Mycology* 2: 1–17.

Schneider, T. R. 2002. A genetic algorithm for the identification of conformationally invariant regions in protein molecules. *Acta Crystallographica Section D: Biological Crystallography* 58: 195–208.

Schneider, T., Gries, K., Josten, M., et al. 2009. The lipopeptide antibiotic Friulimicin B inhibits cell wall biosynthesis through complex formation with bactoprenol phosphate. *Antimicrobial Agents and Chemotherapy* 53: 1610–1618.

Sharma, D., Mandal, S. M. and Manhas, R. K. 2014. Purification and characterization of a novel lipopeptide from *Streptomyces amritsarensis* sp. nov. active against methicillin-resistant Staphylococcus aureus. *AMB Express* 4: 50.

Shoji, J., Hinoo, H., Sakazaki, R., et al. 1978. Isolation of tridecaptins A, B and C (studies on antibiotics from the genus *Bacillus*. XXIII). *Journal of Antibiotics* 31: 646–651.

Shoji, J. I., Hinoo, H. and Sakazaki, R. 1976. The constituent amino acids and fatty acid of antibiotic 333-25. *The Journal of Antibiotics* 29: 521–525.

Shoji, J.-I., Kozuki, S., Okamoto, S., Sakazaki, R. and Otsuka, H. 1968. Studies on tsushimycin. I. *The Journal of Antibiotics* 21: 439–443.

Tanaka, H., Oiwa, R., Matsukura, S., Inokoshi, J. and Omura, S. 1982. Studies on bacterial cell wall inhibitors. *The Journal of Antibiotics* 35: 1216–1221.

Velkov, T., Thompson, P. E., Nation, R. L. and Li, J. 2009. Structure–activity relationships of polymyxin antibiotics. *Journal of Medicinal Chemistry* 53: 1898–1916.

Vértesy, L., Ehlers, E., Kogler, H., et al. 2000. Friulimicins: novel lipopeptide antibiotics with peptidoglycan synthesis inhibiting activity from *Actinoplanes friuliensis* sp. nov. II. Isolation and structural characterization. *The Journal of Antibiotics* 53: 816–827.

Antibacterial Lipopeptides

3.1 POLYMYXIN AS AN ANTIMICROBIAL DRUG

Polymyxins are a class of antibiotic lipopeptides that are produced by non-ribosomal polypeptide synthesis (NRPs). They consist of five cationic polypeptides made of cyclic heptapeptide, a tripeptide that is linear and a tail of fatty acid. This fatty acid is linked to the tripeptide at its N-terminal (Katz and Demain 1977). Seven amino acid residues form the main cyclic component, while another three extend from one of the cyclic residues as a linear chain terminating in either 6-methyloctanoic acid or 6-methylheptanoic acid at the N-terminus. During cyclization, residue 10 is bound to the bridging residue 4 (Dewick 2002). The amino acid residues and diaminobutyric acid (DAB) monomers are generally in the L (levo) configuration, however, certain strains such as *Paenibacillus polymyxa* PKB1 have been observed to incorporate DAB with the D (Dextro) configuration at position 3 producing variations of polymyxin B (Shaheen et al. 2011). The polymyxin family is categorized into several subfamilies, namely, polymyxin A to E, M, S and T based on differences in their fatty acids and amino acid side chains. Polymyxin E, also known as colistin, is further classified as E1 and E2 similar to the classification of polymyxin B as B1 and B2. Polymyxin M is also known as "mattacin" (Martin et al. 2003). Polymyxins are active against a wide range of Gram-negative bacteria, which includes *Acinetobacter baumannii*, *Citrobacter* spp., *Enterobacter* spp., *Escherichia coli*, *Haemophilus influenza*, *Klebsiella* spp.,

Pseudomonas aeruginosa, *Salmonella* spp., *Shigella* spp., and *Yersinia pseudotuberculosis* (Gales et al. 2001, Hogardt et al. 2004, Niks et al. 2004). Polymyxin M and T also exhibit their activity against some Gram-positive bacteria.

3.1.1 Biosynthesis

Similar to other non-ribosomal peptide synthetases (NRPSs), polymyxin synthetases assemble polymyxins by multiple modules, each module contains a set of enzyme domains that proceeds sequentially by recognizing and activating an amino acid residue, adding the next residue leading to the extension of a chain by forming a peptide bond (elongation and condensation reaction) and finally, releasing the peptide chain. Prior to the biosynthesis of polymyxin, the serine residue of peptidyl carrier protein (PCP) is bound to 4'-phosphopantetheine cofactor (ppan). The non-ribosomal peptide synthesis is initiated when an amino acid is activated by domain A of module 1 by a transesterification reaction with ATP to aminoacyl-adenylate. Because of its activating potential, the A domain is designated as adenylation domain. This domain is also required to transfer aminoacyl-adenylate to the free thiol group of PCP-ppan (T-domain). The transfer of amino acid to T-domain leads to the formation of a thioester bond between PCP-ppan and adenylated amino acid. Condensation or C-domain catalyzes the formation of a peptide bond between amino acids. In the final step, involvement of thioesterase (TE) domain at the C-terminal of the last module occurs for the cyclization of molecule and chain is liberated from the enzyme and sent for post-translational modifications (Kopp and Marahiel 2007, Jujjavarapu and Dhagat 2018). Biosynthesis of polymyxin is shown below in Figure 3.1.

3.1.2 Antibacterial Activity of Polymyxin

Polymyxins display a wide spectrum of antibacterial activity. They generally show their antimicrobial activity against Gram-negative bacteria through the disruption of both the inner and outer cell membranes. The important part is its hydrophobic tail, which causes damage to the membrane. This suggests a detergent-like mechanism of action (Velkov et al. 2013). Polymyxins disrupt the outer membrane of bacteria via binding to the lipopolysaccharide. A disruptive physicochemical effect is produced through the binding of polymyxins to phospholipids of the cell wall of Gram-negative bacteria. This leads to changes in the permeability of cell membrane, which ultimately causes cell death (Evans et al. 1999).

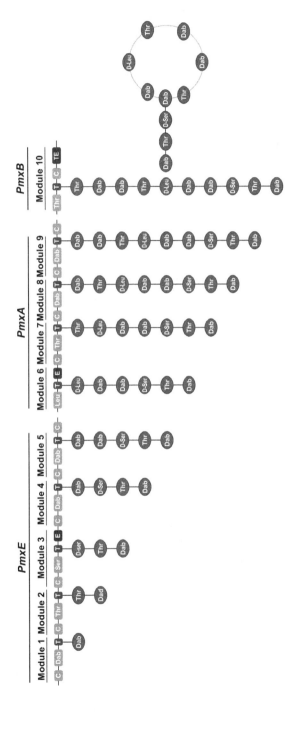

FIGURE 3.1 Biosynthesis of polymyxin.

The initial interaction between polymyxins and the bacterial membranes occurs via electrostatic interactions between the anionic LPS molecule and the cationic polypeptide in the outer membrane of the Gram-negative bacteria. The L-DAB molecule present in polymyxin contains a positive charge, while lipopolysaccharide present on the cell wall of bacteria is negatively charged. Divalent cations such as calcium (Ca^{2+}) and magnesium (Mg^{2+}) (Newton 1956, Davis et al. 1971, Schindler and Osborn 1979) present in the cell membrane of bacteria normally stabilize the LPS molecule, but polymyxins with a L-DAB molecule possess higher affinity toward LPS than the other divalent cations like Mg^{2+} and Ca^{2+}. Polymyxins create a local disturbance from the negatively charged LPS in the outer membrane by displacing magnesium (Mg^{2+}) and the calcium (Ca2+). This process causes an increase in the permeability of the cell envelope, which consists of a cytoplasmic membrane and cell wall. This ultimately leads to cell content leakage and subsequently cell death (Newton 1956, Davis et al. 1971). However, calcium and magnesium in high concentrations can antagonize the interaction of LPS molecule with polymyxins (Davis et al. 1971, Hancock and Chapple 1999). It is suggested that this detergent effect increases the susceptibility of Gram-negative bacteria to other antimicrobial drugs with subsequent exposure of polymyxins. Polymyxin B converts into polymyxin nonapeptide when its hydrophobic tail is removed. It can still bind to lipopolysaccharide but loses the potential to kill bacterial cells. However, the permeability of the cell wall of the bacteria still increases for other antibiotics. This indicates that the polymyxin nonapeptide still retains the ability to cause membrane disorganization to some extent (Tsubery et al. 2000). Resistance can be developed in Gram-negative bacteria against polymyxins via modification in the structure of lipopolysaccharide thus inhibiting the binding of polymyxins to lipopolysaccharide (Tran et al. 2005). Notable cases of increased antibiotic resistance of polymyxins are found specifically in Southern China. In recent studies gene, mcr-1 is isolated from the plasmid of *Enterobacteriaceae,* which is responsible for antibiotic resistance (Wolf 2015, Liu et al. 2016).

3.1.3 Identification of Drug Target Sites

Target identification is a process of identification of ligand corresponding to a drug molecule. A ligand is a small molecule present on the bacterial cell surface onto which the drug of interest binds. This is required for the interaction of a given drug at the target site. The target site is a three-dimensional space inside a target molecule where the interaction of ligand molecule with drug takes place. The first step in *in silico* drug design is the identification of ligands for the respective drug. This is performed by the process of data mining to search for unique and suitable ligands from the available databases. One of the most famous databases available for ligand data mining is RCSB PDB (https://www.rcsb.org/)

Antibiotic resistance is increasing in Gram-negative bacteria, especially in *P. aeruginosa*, *Klebsiella pneumonia* and *A. baumannii*. This has emerged as a global challenge in the current medical world. Due to the dry pipeline of antibiotic discovery, there are no new antibiotics suitable for these "superbugs" in the near future. For the effective treatment of infections caused by multidrug-resistant strains of Gram-negative bacteria, Polymyxin B and colistin have been significantly used as the last therapeutic option (Velkov et al. 2013). The available 3D structure of polymyxin has been retrieved from the protein data bank (PDB) with PDB ID as 2JSO. The selected ligands were chosen on the basis of mode of action of polymyxin, which depicts that DAB is a small molecule present in the lipopolysaccharide of the bacterial cell membrane. DAB is a small ligand that interacts with the positively charged lipopeptide polymyxin to cause the death of the bacterial cell. Type II NADH-quinone oxidoreductase has been chosen as a second ligand which can be targeted by polymyxin. It is a respiratory enzyme that takes part in the oxidation-reduction reaction and is present in the inner membrane of bacteria. This enzyme forms an essential part of the electron transport pathway of bacteria. Current studies depict that polymyxins show inhibitory activity against NADH-quinone oxidoreductase enzyme in both the categories of bacteria, that is, Gram-negative and Gram-positive bacteria (Deris et al. 2014). The list of all possible ligands of polymyxin is summarized in Table 3.1 along with their structures.

3.1.4 Ligand-Based Molecular Docking

Molecular docking is a tool to determine the interaction of the ligand with its corresponding drug molecule. This enables the generation of a site in the 3D cavity of drug molecules where the ligands can bind. The amino acids present in the 3D cavity of drug molecules provide necessary electrons for the formation of hydrogen bonds between the ligand and drug molecule. The number of hydrogen bond is independent of the docking score and glide energy. Docking score determines affinity between the drug and its ligand. The lower the docking score means the higher the affinity between the molecules docked. Glide energy represents the energy

TABLE 3.1   Unique Ligands of Polymyxin and Their 2D Structures

Ligands of Polymyxin	IUPAC Name & Molecular Formula	2D Structure
DAB	2,4-diaminobutyric acid $C_4 H_{10} N_2 O_2$	
NADH	Nicotinamide-Adenine-Dinucleotide	
4FO	(2R)-2,4-diaminobutanoic acid $C_4 H_{10} N_2 O_2$	

of the newly formed complex in the docking process. Lower glide energy represents more stable drug-ligand complex. DAB interacts with polymyxin with the lowest docking score and the lowest glide energy of −4.007 and −24.177 kcal/mol, respectively. This involves the formation of three hydrogen bonds in the complex formed (see Figure 3.2). Similarly, interaction of polymyxin with 4FO resulted in the docking score −4.277 and glide energy −25.213 kcal/mol with four newly formed hydrogen bonds (see Figure 3.3). The results of molecular docking of polymyxin with its ligands are summarized in Table 3.2.

3.1.5 Drug Behavior Analysis Using ADMET

Antibiotic polymyxins are somewhat nephrotoxic and neurotoxic. These are generally used as a last option if the available antibiotics are contraindicated and unsuccessful. These are generally used to treat infections caused by multiple drug-resistant strains of *P. aeruginosa* or carbapenemase-producing Enterobacteriaceae (Velkov et al. 2013). Polymyxins are less effective on Gram-positive microorganisms and are occasionally used in combination with other drugs (such as trimethoprim/polymyxin) to

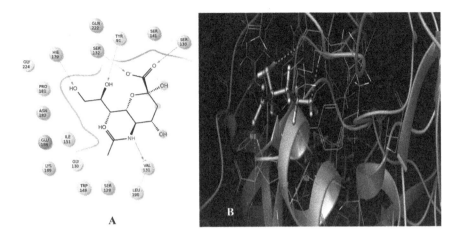

FIGURE 3.2   LID (A) and hydrogen bond interaction (B) of polymyxin and DAB.

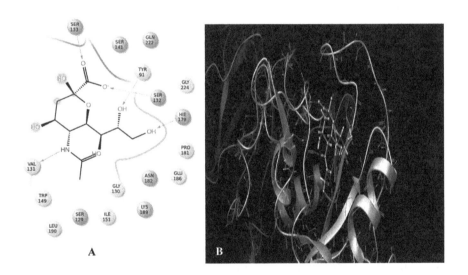

FIGURE 3.3   LID (A) and hydrogen bond interaction (B) of polymyxin and 4FO.

TABLE 3.2   Docking Results of Polymyixn

Ligands	DAB	4FO
H Bond	3	4
Docking Score	−4.007	−4.277
Glide Energy (kcal/mol)	−24.177	−25.213

extend the efficacy (Poirel et al. 2017). Polymyxins cannot be absorbed by the gastrointestinal (GI) tract and hence it is administered orally if it is required to disinfect the GI tract. Other methods for the administration of polymyxin are parenteral (preferably intravenously) and inhalation for systemic treatment (Velkov et al. 2013, Poirel et al. 2017). They are used topically as a drop or cream for the treatment of otitis externa (swimmer's ear) and as a component of triple antibiotic ointment to prevent and treat skin infections (Poirel et al. 2017). The ADMET study is performed via admetSAR, an online tool to predict the molecular descriptors for ADMET. Results obtained using admetSAR are depicted in Table 3.3.

3.1.6 Pharmacophore Models for Polymyxin

Pharmacophore study is conducted to visualize the bond interaction between hydrogen bond acceptor and donor from ligand and protein molecule. Apart from that, the electronegative and electropositive surfaces are also shown in a pharmacophore model. Pharmacophore study, performed here for computer-aided drug design of lipopeptides, is conducted by using an online server named "Pharmit." The software is based upon a previously described defined dataset of molecular descriptors which defines the pharmacophore properties of the query molecule. Pharmacophore models generated for lipopeptides polymyxin with its respective ligand is shown in Figure 3.4. To generate the pharmacophore model, the crystal structure of polymyxin is retrieved from PDB with pdb id 5L3G and ligand 4FO. The respective radius of hydrogen donor and acceptor molecule is shown in Figure 3.4.

3.2 LASPARTOMYCIN

Laspartomycin was initially reported by Umezawa et al. in 1968. It is an acidic lipopeptide antibiotic with a unique peptide core and is structurally related to amphomycin with a 2,3-unsaturated C15–fatty acid side chain (Naganawa et al. 1968). Cyclic lipopeptide laspartomycin C belongs to the family of calcium-dependent antibiotics (CDAs) (Borders et al. 2007). Laspartomycin C is isolated from *Streptomyces viridochromogenes* by fermentation (Naganawa et al. 1968, Borders et al. 2007). It was primarily reported that laspartocin acts against *Staphylococcus aureus in vivo* and *in vitro*, but according to the current studies, laspartomycin C shows its activity against Gram-positive pathogens which include vancomycin-intermediate *S. aureus* (VISA), vancomycin-resistant *S. aureus* (VRSA), vancomycin-resistant Enterococci (VRE) and methicillin-resistant strains

TABLE 3.3   Results of ADMET Study of Polymyixn

ADMET Predicted Profile – Classification		
Model	Result	Probability
Absorption		
Blood–Brain Barrier	BBB-	0.9701
Human Intestinal Absorption	HIA+	0.7115
Caco-2 Permeability	Caco2-	0.7409
P-glycoprotein Substrate	Substrate	0.7282
P-glycoprotein Inhibitor	Non-inhibitor	0.8854
	Non-inhibitor	0.9159
Renal Organic Cation Transporter	Non-inhibitor	0.9525
Distribution		
Subcellular Localization	Mitochondria	0.4481
Metabolism		
CYP450 2C9 Substrate	Non-substrate	0.7661
CYP450 2D6 Substrate	Non-substrate	0.7947
CYP450 3A4 Substrate	Non-substrate	0.5590
CYP450 1A2 Inhibitor	Non-inhibitor	0.8536
CYP450 2C9 Inhibitor	Non-inhibitor	0.9151
CYP450 2D6 Inhibitor	Non-inhibitor	0.9359
CYP450 2C19 Inhibitor	Non-inhibitor	0.8403
CYP450 3A4 Inhibitor	Non-inhibitor	0.8303
CYP Inhibitory Promiscuity	Low CYP Inhibitory Promiscuity	0.9786
Excretion		
Toxicity		
Human Ether-a-go-go-Related	Weak Inhibitor	0.9816
Gene Inhibition	Non-inhibitor	0.8975
AMES Toxicity	Non AMES toxic	0.7723
Carcinogens	Non-carcinogens	0.9274
Fish Toxicity	High FHMT	0.9205
Tetrahymena pyriformis Toxicity	High TPT	0.8952
Honey Bee Toxicity	Low HBT	0.7377
Biodegradation	Not ready biodegradable	0.9002
Acute Oral Toxicity	III	0.6174
Carcinogenicity (Three-class)	Non-required	0.6644
ADMET Predicted Profile – Regression		
Model	Value	Unit
Absorption		
Aqueous Solubility	−2.4789	LogS
Caco-2 Permeability	−0.2181	LogPapp, cm/s

(Continued)

TABLE 3.3 (CONTINUED) Results of ADMET Study of Polymyixn

	Distribution Metabolism Excretion Toxicity	
Rat Acute Toxicity	2.6606	LD50, mol/kg
Fish Toxicity	1.8703	pLC50, mg/L
Tetrahymena pyriformis Toxicity	0.0892	pIGC50, ug/L

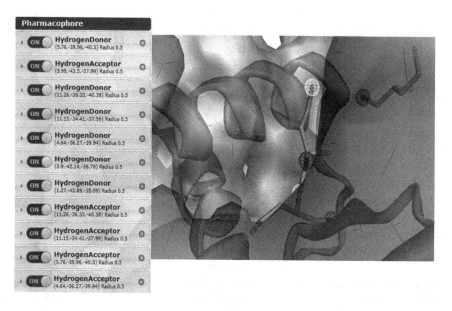

FIGURE 3.4 Pharmacophore model for polymyxin.

of *S. aureus* (MESA) (Naganawa et al. 1968, Curran et al. 2007). The structure of laspartomycin C was first elucidated in 2003 (Kong and Carter 2003, Borders et al. 2007, Yin et al. 2015). It is also known as glycinocin A, which consists of a cyclic core of ten amino acids and an exocyclic region at N-terminal (Kong and Carter 2003). The macrocycle is composed of numerous non-proteinogenic and D-amino acids, which include D-pipecolic acid (D-Pip), L-2,3- diaminopropionic acid (L-2,3-Dap) as well as D-allo-threonine and Asp-XAsp-Gly motif, which is found to be implicated in calcium ion (Ca^{2+}) binding and conserved among the all known CDAs. Laspartomycin consists of a branched and unsaturated C15 fatty acid tail, which is linked to the aspartic acid at the exocyclic N-terminal. The macrocycle of laspartomycin C is closed through an amide linkage, which is found between the proline of C-terminal and position 2 of the side chain of L-2, 3-Dap residue (Kleijn et al. 2016).

3.2.1 Antimicrobial Activity of Laspartomycin

Bacterial resistance toward the conventional antibiotics has emerged as a new challenge which is implicating the growing attention toward the use of CDAs in the past two decades. The increased interest in using CDAs has been driven by the findings which indicate that different CDAs have diversified mode of action unlike the actions of conventional antibiotics (Strieker and Marahiel 2009, Robbel and Marahiel 2010). According to the recent studies, it is found that laspartomycin C has an ability to form a high-affinity complex with the precursor undecaprenyl phosphate (C55-P) present in bacterial cell wall which possesses the capability to inhibit the biosynthesis of peptidoglycan. This inhibition ultimately leads to cell death (Lohani et al. 2015). The mechanism of C55-P has not been exploited by current antibiotics which make it an attractive target to study further. The experimental studies have revealed that the antibacterial activity of laspartomycin increases significantly with the higher concentration of calcium (Ca^{2+}) (Kleijn et al. 2016). The structural insights of these antibiotics are of great value to the design of the antibiotics which have the capability to exploit a unique bacterial target. The incorporation of D-Pip and Pro residues leads to an unfavorable effect which results in subsequent steps of membrane association in the biosynthesis of bacterial cell wall. Further, it is indicated that C55-P is selectively targeted by laspartomycin. Sequestration of C55-P has a major role in blocking the lipid II formation and prevents the biosynthesis of bacterial cell wall. The unique mode of action of laspartomycin C by forming strong complex with C55-P exerts its antibiotic activity. Currently used clinical antibiotics do not possess the mode of action which targets C55-P. Such compounds are of great value in the era of current drug discovery, but further studies are required to validate the therapeutic potential of lipopeptide-based drugs targeting C55-P molecule.

3.2.2 Ligands of Laspartomycin

Key to addressing the growing threat of drug-resistant bacteria is the identification and characterization of antibiotics that operate by unconventional mechanisms (Tommasi et al. 2015, Chellat and Riedl 2017). Notable in this regard are CDAs which have gained clinical prominence owing to their activity against multi-drug-resistant pathogens (Strieker and Marahiel 2009, Schneider et al. 2014, Kleijn et al. 2016). While a variety of antibacterial mechanisms have been ascribed to the various CDAs, an atomic-level understanding of the recognition of their bacterial target(s) remains elusive. The data of unique ligands for laspartomycin have been retrieved from PDB. PDB contains the experimental data associated with the protein which includes the information of unique ligands. Ligands with their two-dimensional structure, molecular formula and IUPAC (International Union of Pure and Applied Chemistry) name for laspartomycin is described in Table 3.4.

3.2.3 Molecular Docking as a Tool for Drug Discovery

A protein molecule with PDB entry 5O0Z was retrieved from RCSB to perform docking. Due to the linear structure of the laspartomycin molecule, groove was not generated for the ligand to bind.

3.2.4 ADMET Properties of Laspartomycin

ADMET properties of laspartomycin are predicted by *in silico* approach using a promising web-based software admetSAR. There is not much experimental data available for the pharmacokinetics of laspartomycin, hence as part of drug discovery, estimation of ADMET property is an indispensable step. The data generated by computational analysis of ADMET properties for laspartomycin is shown in Table 3.5.

3.2.5 Pharmacophore Modeling of Laspartomycin

Lipopeptide laspartomycin is reported in PDB with entry 5O0Z. The molecule 5O0Z is retrieved from the RCSB website (https://www.rcsb.org/structure/5o0z) and ligand 9GB was chosen to perform the pharmacophore study. The result of the pharmacophore study is obtained through "Pharmit" software, which is a web-based tool used to predict the pharmacophore property of a molecule. Results are shown in Figure 3.5. The predicted properties of surface and molecules participating in the drug-ligand interaction with radius is shown in Figure 3.5.

3.3 VANCOMYCIN

Among the currently available antibiotics, vancomycin is structurally unrelated and is a unique glycopeptide. It was first isolated by Edmund Kornfeld (working at Eli Lilly) in 1953 from a soil sample collected from Borneo's interior jungles by a missionary (Shnayerson and Plotkin 2002). The organism found to produce vancomycin is now known as *Amycolatopsis*

TABLE 3.4 Ligands of Laspartomycin with 2D Structures

Ligand	IUPAC Name & Molecular Formula	2D Structure
2TL	D-allothreonine $C_4 H_9 N O_3$	
9GB	[(~{E})-3-methylhex-2-enyl] dihydrogen phosphate $C_7 H_{15} O_4 P$	
9GE	(~{E})-13-methyltetradec-2-enoic acid $C_{15} H_{28} O_2$	
ACY	Acetic acid $C_2 H_4 O_2$	
CA	Calcium ion Ca	Ca^{2+}
CL	Chloride ion Cl^-	Cl^-
CPI	6-Carboxypiperidine $C_6 H_{11} N O_2$	
DNP	3-Amino-alanine $C_3 H_9 N_2 O_2$	

TABLE 3.5   ADMET Results of Laspartomycin

ADMET Predicted Profile – Classification		
Model	**Result**	**Probability**
Absorption		
Blood–Brain Barrier	BBB-	0.9830
Human Intestinal Absorption	HIA+	0.9334
Caco-2 Permeability	Caco2-	0.7550
P-glycoprotein Substrate	Substrate	0.8090
P-glycoprotein Inhibitor	Non-inhibitor	0.7688
	Non-inhibitor	0.9245
Renal Organic Cation Transporter	Non-inhibitor	0.9386
Distribution		
Subcellular Localization	Mitochondria	0.5166
Metabolism		
CYP450 2C9 Substrate	Non-substrate	0.8522
CYP450 2D6 Substrate	Non-substrate	0.7623
CYP450 3A4 Substrate	Substrate	0.5614
CYP450 1A2 Inhibitor	Non-inhibitor	0.9192
CYP450 2C9 Inhibitor	Non-inhibitor	0.8276
CYP450 2D6 Inhibitor	Non-inhibitor	0.8702
CYP450 2C19 Inhibitor	Non-inhibitor	0.7647
CYP450 3A4 Inhibitor	Non-inhibitor	0.6078
CYP Inhibitory Promiscuity	Low CYP Inhibitory Promiscuity	0.9130
Excretion		
Toxicity		
Human Ether-a-go-go-Related Gene Inhibition	Weak Inhibitor	0.9417
	Non-inhibitor	0.5909
AMES Toxicity	Non AMES toxic	0.7891
Carcinogens	Non-carcinogens	0.7762
Fish Toxicity	High FHMT	0.9865
Tetrahymena pyriformis Toxicity	High TPT	0.9378
Honey Bee Toxicity	Low HBT	0.7309
Biodegradation	Not ready biodegradable	0.9433
Acute Oral Toxicity	I	0.7517
Carcinogenicity (Three-class)	Non-required	0.6348
ADMET Predicted Profile – Regression		
Model	**Value**	**Unit**
Absorption		
Aqueous Solubility	−2.9678	LogS
Caco-2 Permeability	−0.0480	LogPapp, cm/s
		(*Continued*)

TABLE 3.5 (CONTINUED)   ADMET Results of Laspartomycin

	Distribution Metabolism Excretion Toxicity	
Rat Acute Toxicity	4.2302	LD50, mol/kg
Fish Toxicity	1.5911	pLC50, mg/L
Tetrahymena pyriformis Toxicity	0.2683	pIGC50, ug/L

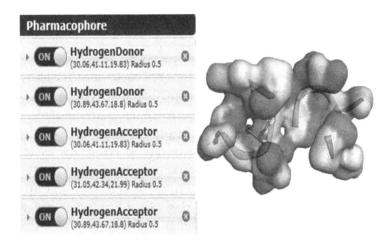

FIGURE 3.5   Pharmacophore model for laspartomycin.

orientalis (Levine 2006). Originally, it was indicated that vancomycin was used to treat the infections caused by penicillin-resistant *S. aureus* (Levine 2006, Moellering Jr 2006). It is a tricyclic branched glycosylated non-ribosomal peptide which is produced by fermentation from bacterial species of *Actinobacteria*, i.e. *A. orientalis*, a soil bacterium formerly known as *Nocardia orientalis* (Villa et al. 2016). Vancomycin is often used as a last option when other antibiotics fail and hence it is called the "drug of last resort." It is generally used to treat various bacterial infections of Gram-positive bacteria which are life-threatening, serious and insensitive to other antibiotics. Vancomycin shows its activity in clinical infections and *in vitro* against the following strains of microorganisms: *Streptococcus pyrogens*, *S. agalactiae*, *S. pneumonia* (including penicillin-resistant strains), *Listeria monocytogens*, *Lactobacillus* species and *Actinomyces* species. It is recommended to administer it intravenously for the treatment of complicated infections in skin, bloodstream, endocarditis, joint, bone and meningitis which is caused by a methicillin-resistant strain of *S. aureus* (Liu et al. 2011). The antibacterial drug vancomycin is also obtained by *Streptomyces orientalis*. This is related to ristocetin that inhibits the assembly of the bacterial cell wall and possesses a toxic effect on the inner ear and the kidney. CutisPharma has FDA approval for vancomycin, named Firvanq, from 29 January 2018. Firvanq, in the form of an oral liquid, is the only current FDA approved treatment option of vancomycin, which is used to treat diarrhea caused by *Clostridium difficile* and Enterocolitis associated with *S. aureus*. The treatment also covers methicillin-resistant strains [LP1196] (https://www.drugbank.ca/drugs/DB00512).

3.3.1 Biosynthesis of Vancomycin

Vancomycin is produced by *A. orientalis*, a soil bacterium. The biosynthesis of vancomycin occurs via various non-ribosomal peptide synthetases (NRPSs) (Samel et al. 2008). During the assembly of amino acids, NRPS enzymes determine the sequence of the peptide by seven modules. The amino acids first undergo certain modifications before the assembling of vancomycin via NRPS. The modification of L-tyrosine converts it into 4-hydroxyphenylglycine (HPG) and β-hydroxy chlorotyrosine (β-hTyr) residues. The derivation of 3,5-DPG ring (3,5-dihydroxyphenylglycine) occurs by the use of acetate (Dewick 2002). The distinct modules of non-ribosomal peptide synthesis consist of loading and extension of protein by the addition of each amino acid through the formation of an amide bond. The amide bonds are formed in the activating domains at their contact sites (van Wageningen et al. 1998). Every module consists of a domain of adenylation (A), a domain of PCP and the domains of condensation (C) and elongation (E). Domain A includes the process of thioesterification in which the activation of specific amino acid is done via conversion into an enzyme complex of aminoacyl adenylate with 4′-phosphopantetheine cofactor (Schlumbohm et al. 1991, Stein et al. 1996). This complex is transferred further to the PCP domain by exclusion of AMP. The attached prosthetic group, 4′-phosphopantethein is used by PCP domain for loading of their precursors (two) and the rising peptide chain (Kohli et al. 2002).

Certain additional domains of modification are also present for the biosynthesis of vancomycin, that is, epimerization (E). This domain takes part in the isomerization of amino acids, which alters the stereochemistry from one isomeric form to another form. A thioesterase (TE) domain catalyzes the cyclization and is used to release the molecule through a TE scission. The assembly of heptapeptide is carried out by a set of multienzymes, namely, peptide synthetases CepA, CepB and CepC. The association of CepA, CepB and CepC resembles the peptide synthetases, for example, gramicidin (GrsA and GrsB) and surfactin (SrfA1, SrfA2 and SrfA3) (van Wageningen et al. 1998). The activation of each domain is sequestered by each peptide synthetases for activating the codes for amino acids. Module 1, 2 and 3 are coded by CepA, module 4, 5 and 6 are coded by CepB and further coding of module 7 is done by CepC. The locations of three peptide synthetases in the bacterial genome are found at the start region, which is linked with the antibiotic biosynthesis. The length of this region is 27 kb (van Wageningen et al. 1998). Once the molecule of linear heptapeptide is synthesized, vancomycin goes through further modifications, such as glycosylation and oxidative cross-linking, to develop as a biologically active molecule.

3.3.2 Action of Vancomycin against Bacteria

Vancomycin inhibits the induction of bacterial L-phase in prone microorganisms. The unique mode of action of vancomycin creates a hindrance for susceptible bacteria by inhibiting the cell wall synthesis at its second stage. Vancomycin alters the permeability of bacterial cell membrane and selectively inhibits the synthesis of ribonucleic acid. It inhibits the biosynthesis of bacterial cell wall. Vancomycin exhibits its mechanism of action based upon its capability to form complex with moieties of D-alanyl-D-alanine present in the polymer synthesis phase (Hammes and Neuhaus 1974). The primary result of bactericidal action starts by preventing the incorporation of N-acetylglucosamine (NAG) and N-acetylmuramic acid (NAM) peptide subunits into the matrix of peptidoglycan. These peptide subunits are the major structural components of the cell wall of Gram-positive bacteria. Vancomycin contains a large hydrophilic molecule which has the ability to form hydrogen bond interaction with D-ala-D-ala moiety of N-acetyl muramic acid and N-acetylglucosamine-peptide, also known as NAM/NAG-peptide subunit. This interaction occurs at five points. The interaction of vancomycin to D-ala-D-ala leads to the prevention of subunits of NAM/NAG-peptide being incorporated into the matrix of peptidoglycan. This exhibits the alteration in RNA synthesis and the permeability of cell membrane in bacteria. No cross-resistance has been notified in other antibiotics and vancomycin. Resistance to vancomycin has not been increased in the last three decades. An *in vitro* synergistic act of increased antimicrobial activity of vancomycin has been reported with the combination of the aminoglycoside. Such enhanced antimicrobial activity has been demonstrated against *Streptococcus bovis*, *S. aureus*, *Streptococcus viridians* and Enterococci. The combination of vancomycin with rifampicin is antagonistic to various strains of *S. aureus*, though the effect is occasionally synergistic and indifferent. This combination shows synergism against strains of *Staphylococcus epidermidis* and indifference against Enterococci. The combination of vancomycin with fusidic acid does not make any difference against *S. aureus*. Vancomycin does not act upon Gram-negative *Bacilli*, fungi or mycobacteria. (https://www.drug-bank.ca/drugs/DB00512)

3.3.3 Ligand Identification of Vancomycin

Ligand identification of lipopeptide drug vancomycin is done using the universal protein repository of RCSB known as PDB. PDB provides the information about available unique ligands of a given drug. Unique ligands are the small compounds which have specific interaction with the drug molecule. These ligands present on the cell wall of bacteria provide a site for the interaction to inhibit the growth of bacteria by their specific mode of actions, discussed previously. Here, the 3D structure of vancomycin is retrieved in a .pdb format with PDB ID: 1AA5. The universal protein repository, PDB, provides the information about nine unique ligands for vancomycin which can be retrieved from the website https://www.rcsb. org/. The ligands are listed in Table 3.6. The 2D structure of each ligand has been drawn using a software-based tool named 2D Sketcher, from Schrödinger software.

3.3.4 Studies on Molecular Docking of Vancomycin

Molecular docking studies of vancomycin were performed with its unique ligands retrieved from the PDB. Docking performed with each ligand shows the results which contain the information about the number of hydrogen bonds formed between the given ligand and the drug molecule. Table 3.7 depicts the information about the number of hydrogen bonds formed between the drug and the ligand, docking score and glide energy generated during docking.

Docking was performed by taking the 3D structure of each ligand and the drug vancomycin which is used in .pdb format retrieved from the PDB. Ligand 3FG has the interaction of two hydrogen bonds with vancomycin and glide energy generated is –27.147 kcal/mol. The docking score generated as a result of interaction between 3FG and vancomycin is –4.820 which is the lowest among all the nine unique ligands used for

TABLE 3.6   Ligands Name of Vancomycin and Their 2D Structures

Ligands	IUPAC Name & Molecular Formula	2D Structure
ACY	Acetic acid $C_2 H_4 O_2$	
BGC	β-D-Glucose $C_6 H_{12} O_6$	
CL	Chloride ion Cl⁻	Cl⁻
GHP	(2R)-amino (4-hydroxyphenyl) ethanoic acid $C_8 H_9 N O_3$	
MLU	N-methyl-D-leucine $C_7 H_{15} N O_2$	

(*Continued*)

TABLE 3.6 (CONTINUED) Ligands Name of Vancomycin and Their 2D Structures

Ligands	IUPAC Name & Molecular Formula	2D Structure
OMY	(βR)-3-chloro-β-hydroxy-L-tyrosine $C_9 H_{10} Cl N O_4$	
OMZ	(βR)-3-chloro-β-hydroxy-D-tyrosine $C_9 H_{10} Cl N O_4$	
RER	(1R,3S,4S,5S)-3-amino-2,3,6-trideoxy-3-methyl- α-L-arabino-hexopyranose Vancosamine (Synonym) $C_7 H_{15} N O_3$	

TABLE 3.7 Docking Results of Vancomycin

Unique Ligands	3FG	BGC	MLU	GHP	RER	OMY	OMZ
H Bond	2	1	2	1	2	2	1
Docking Score	−4.820	−4.230	−3.486	−3.412	−4.603	−3.706	−3.348
Glide Energy (kcal/mol)	−27.147	−20.605	−19.065	−20.837	−18.031	−28.338	−25.257

docking (see Figure 3.6). The ligand BGC interacts with vancomycin with one hydrogen bond and −4.23 and −20.605 kcal/mol of docking score and glide energy, respectively (see Figure 3.7). Similarly, MLU interacted with the molecule with two hydrogen bonds with docking score of −3.486 and

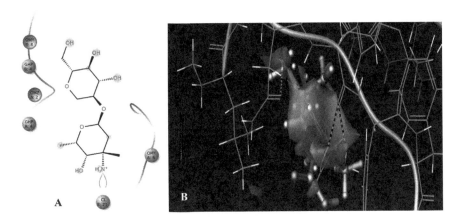

FIGURE 3.6　LID (A) and hydrogen bond interaction (B) between vancomycin and 3FG.

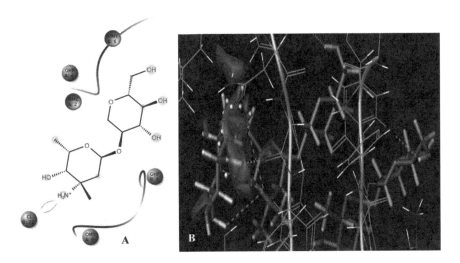

FIGURE 3.7　LID (A) and hydrogen bond interaction (B) between vancomycin and BGC.

glide energy of −19.065 kcal/mol (see Figure 3.8). Another ligand GHP had docking score and glide energy of −3.412 and −20.837, respectively with one hydrogen bond (see Figure 3.9) whereas RER interacted by forming two hydrogen bonds with glide energy and docking score of −18.031 kcal/mol and −4.603, respectively (see Figure 3.10). Also, OMY formed two hydrogen bonds with vancomycin and had a docking score of −3.706

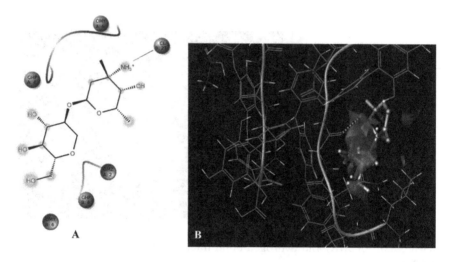

FIGURE 3.8 LID (A) and hydrogen bond interaction (B) between vancomycin and MLU.

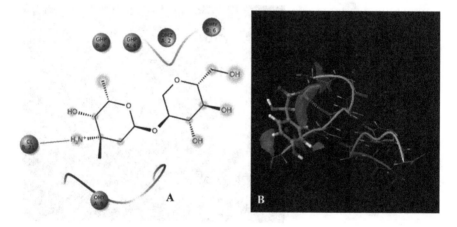

FIGURE 3.9 LID (A) and hydrogen bond interaction (B) between vancomycin and GHP.

and glide energy of −28.338 kcal/mol (see Figure 3.11). Lastly, OMZ formed one hydrogen bond with the drug with docking score of −3.348 and glide energy of −25.257 kcal/mol (see Figure 3.12). Here, as per the results, ligand 3FG has the best interaction with vancomycin. The ligand interaction diagram (LID) and the images consisting of hydrogen bond interaction between the ligand and vancomycin are shown as dotted lines. Chlorine ion and acetic acid (ACY) were not considered as ideal ligands for docking because they are very small molecules with respect to the drug used for docking.

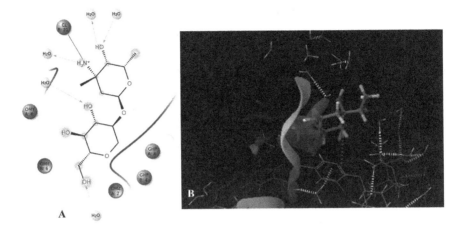

FIGURE 3.10 LID (A) and hydrogen bond interaction (B) between vancomycin and RER.

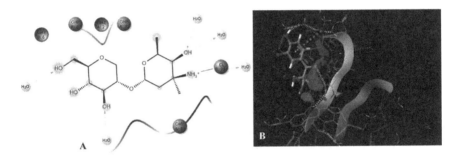

FIGURE 3.11 LID (A) and hydrogen bond interaction (B) between vancomycin and OMY.

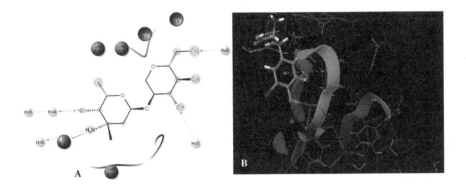

FIGURE 3.12 LID (A) and hydrogen bond interaction (B) between vancomycin and OMZ.

3.3.5 ADMET Studies of Vancomycin

ADMET study of vancomycin is performed using an online tool named admetSAR (http://lmmd.ecust.edu.cn/admetsar1/predict/). The assessment is based upon a certain knowledge-based set of rules. These knowledge-based rules include checking the number of metabolites, suitable percentage values of human oral absorption, logP, number of rotatable bonds, solubility and cell permeability. The results obtained using the tool are shown in Table 3.8. The predicted brain/blood partition coefficient for an orally delivered form of vancomycin is 0.9910

TABLE 3.8 ADMET Results of Vancomycin

ADMET Predicted Profile – Classification		
Model	**Result**	**Probability**
Absorption		
Blood–Brain Barrier	BBB-	0.9910
Human Intestinal Absorption	HIA-	0.7876
Caco-2 Permeability	Caco2-	0.7094
P-glycoprotein Substrate	Substrate	0.8562
P-glycoprotein Inhibitor	Non-inhibitor	0.8781
	Non-inhibitor	0.9636
Renal Organic Cation Transporter	Non-inhibitor	0.9503
Distribution		
Subcellular Localization	Lysosome	0.5110
Metabolism		
CYP450 2C9 Substrate	Non-substrate	0.8535
CYP450 2D6 Substrate	Non-substrate	0.8323
CYP450 3A4 Substrate	Substrate	0.6686
CYP450 1A2 Inhibitor	Non-inhibitor	0.9046
CYP450 2C9 Inhibitor	Non-inhibitor	0.9071
CYP450 2D6 Inhibitor	Non-inhibitor	0.9231
CYP450 2C19 Inhibitor	Non-inhibitor	0.9026
CYP450 3A4 Inhibitor	Non-inhibitor	0.8309
CYP Inhibitory Promiscuity	Low CYP Inhibitory Promiscuity	0.7268
Excretion		
Toxicity		
Human Ether-a-go-go-Related Gene Inhibition	Weak Inhibitor	0.9987
	Non-inhibitor	0.8098
AMES Toxicity	Non AMES toxic	0.5927
		(*Continued*)

TABLE 3.8 (CONTINUED)   ADMET Results of Vancomycin

Carcinogens	Non-carcinogens	0.8826
Fish Toxicity	High FHMT	0.9099
Tetrahymena pyriformis Toxicity	High TPT	0.9932
Honey Bee Toxicity	Low HBT	0.7402
Biodegradation	Not ready biodegradable	1.0000
Acute Oral Toxicity	III	0.6193
Carcinogenicity (Three-class)	Non-required	0.4997
ADMET Predicted Profile – Regression		
Model	**Value**	**Unit**
	Absorption	
Aqueous Solubility	−3.4440	LogS
Caco-2 Permeability	0.0397	LogPapp, cm/s
	Distribution	
	Metabolism	
	Excretion	
	Toxicity	
Rat Acute Toxicity	2.5856	LD50, mol/kg
Fish Toxicity	1.2076	pLC50, mg/L
Tetrahymena pyriformis Toxicity	0.6265	pIGC50, ug/L

which is under the acceptable range. Vancomycin is recommended to be taken by mouth as a treatment for severe *C. difficile* colitis but the oral absorption is very poor. The probability of human intestinal absorption is 0.7876 which is less than 1, which depicts the poor absorption of the drug in the intestine.

3.3.6 Ligand-Based Pharmacophore Modeling of Vancomycin

A molecular model describing the spatial arrangements of hydrogen donor and acceptor molecules with the properties of molecular surface and the spatial arrangements of atoms of interaction is known as a pharmacophore model. The pharmacophore property describing the stereochemistry of vancomycin is studied using the online tool "pharmit." The vancomycin structure is retrieved from PDB from RCSB with pdb entry 1AA5. The pharmacophore model is generated for the interaction of 1AA5 with its unique ligand 3FG. The pharmacophore model for vancomycin is shown in Figure 3.13.

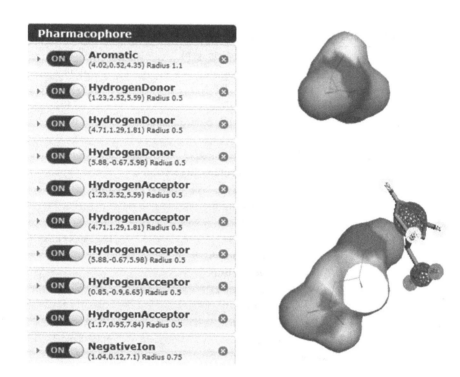

FIGURE 3.13   Pharmacophore model for vancomycin

3.4 SUMMARY

Polymyxin and vancomycin are both generally preferred as antibiotics of last resort. These are used to treat infections caused by multidrug-resistant microorganisms. Polymyxin is specifically used to treat Gram-negative bacteria while vancomycin attacks the Gram-positive bacteria. The mechanism of action of polymyxin consists of its detergent-like property wherein the hydrophobic tail disrupts the cell wall of Gram-negative bacteria. On the other hand, vancomycin disturbs the incorporation of NAM and NAG subunits into the cell wall peptidoglycan of Gram-positive bacteria. According to the pharmacokinetic analysis of these drugs, it is found that polymyxin is not suitable for oral administration and hence it is generally used for tropical applications. In certain cases, such as to disinfect the GI tract, polymyxin is used orally as it cannot be absorbed by the GI tract. Vancomycin results in poor oral absorption; therefore, it is suggested for intravenous administration. Because of their poor absorption, vancomycin and polymyxin exhibit neurotoxicity and nephrotoxicity but are suitable to combat severe infections. Hence, they are used as a last option against antibiotic-resistant bacteria. Laspartomycin comes under the –category of calcium-dependent antibiotics. It is generally used to treat Gram-positive bacteria. Laspartomycin targets the C55-P molecule to inhibit the bacterial cell wall biosynthesis. It is clinically important because of its specificity in treating microorganisms resistant against vancomycin and methicillin-mediated *S. aureus*. The treatment which contains low susceptibility toward vancomycin can be treated by laspartomycin. Molecular docking studies of vancomycin and polymyxin also reveal better results as compared to other lipopetipdes described in other chapters, hence proving the better drug-gability via *in silico* drug design methods.

REFERENCES

Borders, D. B., Leese, R. A., Jarolmen, H., et al. 2007. Laspartomycin, an acidic lipopeptide antibiotic with a unique peptide core. *Journal of Natural Products* 70: 443–446.

Chellat, M. F. and Riedl, R. 2017. Pseudouridimycin: the first nucleoside analogue that selectively inhibits bacterial RNA polymerase. *Angewandte Chemie International Edition* 56: 13184–13186.

Curran, W. V., Leese, R. A., Jarolmen, H., et al. 2007. Semisynthetic approaches to laspartomycin analogues. *Journal of Natural Products* 70: 447–450.

Davis, S. D., Iannetta, A. and Wedgwood, R. J. 1971. Activity of colistin against *Pseudomonas aeruginosa*: inhibition by calcium. *Journal of Infectious Diseases* 124: 610–612.

Deris, Z. Z., Akter, J., Sivanesan, S., et al. 2014. A secondary mode of action of polymyxins against Gram-negative bacteria involves the inhibition of NADH-quinone oxidoreductase activity. *The Journal of Antibiotics* 67: 147.

Dewick, P. M. 2002. *Medicinal Natural Products: A Biosynthetic Approach*, John Wiley &Sons.

Evans, M. E., Feola, D. J. and Rapp, R. P. 1999. Polymyxin B sulfate and colistin: old antibiotics for emerging multiresistant gram-negative bacteria. *Annals of Pharmacotherapy* 33: 960–967.

Gales, A. C., Reis, A. O. and Jones, R. N. 2001. Contemporary assessment of antimicrobial susceptibility testing methods for polymyxin B and colistin: review of available interpretative criteria and quality control guidelines. *Journal of Clinical Microbiology* 39: 183–190.

Hammes, W. P. and Neuhaus, F. C. 1974. On the mechanism of action of vancomy-cin: inhibition of peptidoglycan synthesis in *Gaffkya homari*. *Antimicrobial Agents and Chemotherapy* 6: 722–728.

Hancock, R. E. and Chapple, D. S. 1999. Peptide antibiotics. *Antimicrobial Agents and Chemotherapy* 43: 1317–1323.

Hogardt, M., Schmoldt, S., Götzfried, M., Adler, K. and Heesemann, J. 2004. Pitfalls of polymyxin antimicrobial susceptibility testing of *Pseudomonas aeruginosa* isolated from cystic fibrosis patients. *Journal of Antimicrobial Chemotherapy* 54: 1057–1061.

Jujjavarapu, S. E. and Dhagat, S. 2018. In silico discovery of novel ligands for antimicrobial lipopeptides for computer-aided drug design. *Probiotics and Antimicrobial Proteins* 10: 129–141.

Katz, E. and Demain, A. L. 1977. The peptide antibiotics of *Bacillus*: chemistry, biogenesis, and possible functions. *Bacteriological Reviews* 41: 449.

Kleijn, L. H., Oppedijk, S. F., 't Hart, P., et al. 2016. Total synthesis of laspartomycin C and characterization of its antibacterial mechanism of action. *Journal of Medicinal Chemistry* 59: 3569–3574.

Kohli, R. M., Walsh, C. T. and Burkart, M. D. 2002. Biomimetic synthesis and optimization of cyclic peptide antibiotics. *Nature* 418: 658.

Kong, F. and Carter, G. T. 2003. Structure determination of glycinocins A to D, further evidence for the cyclic structure of the amphomycin antibiotics. *The Journal of Antibiotics* 56: 557–564.

Kopp, F. and Marahiel, M. A. 2007. Macrocyclization strategies in polyketide and nonribosomal peptide biosynthesis. *Natural Product Reports* 24: 735–749.

Levine, D. P. 2006. Vancomycin: a history. *Clinical Infectious Diseases* 42: S5–S12.

Liu, C., Bayer, A., Cosgrove, S. E., et al. 2011. Clinical practice guidelines by the Infectious Diseases Society of America for the treatment of methicillin-resistant *Staphylococcus aureus* infections in adults and children. *Clinical Infectious Diseases* 52: e18–e55.

Liu, Y.-Y., Wang, Y., Walsh, T. R., et al. 2016. Emergence of plasmid-mediated colistin resistance mechanism MCR-1 in animals and human beings in China: a microbiological and molecular biological study. *The Lancet Infectious Diseases* 16: 161–168.

Lohani, C. R., Taylor, R., Palmer, M. and Taylor, S. D. 2015. Solid-phase total synthesis of daptomycin and analogs. *Organic Letters* 17: 748–751.

Martin, N. I., Hu, H., Moake, M. M., et al. 2003. Isolation, structural characterization, and properties of mattacin (polymyxin M), a cyclic peptide antibiotic produced by *Paenibacillus kobensis* M. *Journal of Biological Chemistry* 278: 13124–13132.

Moellering Jr, R. C. 2006. *Vancomycin: A 50-Year Reassessment*, The University of Chicago Press, Chicago, IL.

Naganawa, H., Hamada, M., Maeda, K., et al. 1968. Laspartomycin, a new anti-staphylococcal peptide. *The Journal of Antibiotics* 21: 55–62.

Newton, B. 1956. The properties and mode of action of the polymyxins. *Bacteriological Reviews* 20: 14.

Niks, M., Hanzen, J., Ohlasová, D., et al. 2004. Multiresistant nosocomial bacterial strains and their "in vitro" susceptibility to chloramphenicol and colistin. *Klinicka mikrobiologie a infekcni lekarstvi* 10: 124–129.

Poirel, L., Jayol, A. and Nordmann, P. 2017. Polymyxins: antibacterial activity, susceptibility testing, and resistance mechanisms encoded by plasmids or chromosomes. *Clinical Microbiology Reviews* 30: 557–596.

Robbel, L. and Marahiel, M. A. 2010. Daptomycin, a bacterial lipopeptide synthesized by a nonribosomal machinery. *Journal of Biological Chemistry* 285: 27501–27508.

Samel, S. A., Marahiel, M. A. and Essen, L.-O. 2008. How to tailor non-ribosomal peptide products—new clues about the structures and mechanisms of modifying enzymes. *Molecular Biosystems* 4: 387–393.

Schindler, M. and Osborn, M. 1979. Interaction of divalent cations and polymyxin B with lipopolysaccharide. *Biochemistry* 18: 4425–4430.

Schlumbohm, W., Stein, T., Ullrich, C., et al. 1991. An active serine is involved in covalent substrate amino acid binding at each reaction center of gramicidin S synthetase. *Journal of Biological Chemistry* 266: 23135–23141.

Schneider, T., Müller, A., Miess, H. and Gross, H. 2014. Cyclic lipopeptides as antibacterial agents–potent antibiotic activity mediated by intriguing mode of actions. *International Journal of Medical Microbiology* 304: 37–43.

Shaheen, M., Li, J., Ross, A. C., Vederas, J. C. and Jensen, S. E. 2011. *Paenibacillus polymyxa* PKB1 produces variants of polymyxin B-type antibiotics. *Chemistry & Biology* 18: 1640–1648.

Shnayerson, M. and Plotkin, M. J. 2002. *The Killers Within: The Deadly Rise Of Drug-Resistant Bacteria*, Little, Brown and Company, New York.

Stein, T., Vater, J., Kruft, V., et al. 1996. The multiple carrier model of nonribosomal peptide biosynthesis at modular multienzymatic templates. *Journal of Biological Chemistry* 271: 15428–15435.

Strieker, M. and Marahiel, M. A. 2009. The structural diversity of acidic lipopeptide antibiotics. *ChemBioChem* 10: 607–616.

Tommasi, R., Brown, D. G., Walkup, G. K., Manchester, J. I. and Miller, A. A. 2015. ESKAPEing the labyrinth of antibacterial discovery. *Nature Reviews Drug Discovery* 14: 529.

Tran, A. X., Lester, M. E., Stead, C. M., et al. 2005. Resistance to the antimicrobial peptide polymyxin requires myristoylation of *Escherichia coli* and *Salmonella typhimurium* lipid A. *Journal of Biological Chemistry* 280: 28186–28194.

Tsubery, H., Ofek, I., Cohen, S. and Fridkin, M. 2000. Structure–function studies of polymyxin B nonapeptide: implications to sensitization of gram-negative bacteria. *Journal of Medicinal Chemistry* 43: 3085–3092.

van Wageningen, A. A., Kirkpatrick, P. N., Williams, D. H., et al. 1998. Sequencing and analysis of genes involved in the biosynthesis of a vancomycin group antibiotic. *Chemistry &Biology* 5: 155–162.

Velkov, T., Roberts, K. D., Nation, R. L., Thompson, P. E. and Li, J. 2013. Pharmacology of polymyxins: new insights into an 'old' class of antibiotics. *Future Microbiology* 8: 711–724.

Villa T.G., Feijoo-Siota L., Rama J.L.R., Sánchez-Pérez A., de Miguel-Bouzas T. 2016. The Case of Lipid II: The Achilles' Heel of Bacteria. In: Villa T., Vinas M. (eds) *New Weapons to Control Bacterial Growth*. Springer, Cham, Switzerland.

Wolf, J. 2015. Antibiotic resistance threatens the efficacy of prophylaxis. *The Lancet Infectious Diseases* 15: 1368–1369.

Yin, N., Li, J., He, Y., et al. 2015. Structure–activity relationship studies of a series of semisynthetic lipopeptides leading to the discovery of Surotomycin, a novel cyclic lipopeptide being developed for the treatment of *Clostridium difficile*-associated diarrhea. *Journal of Medicinal Chemistry* 58: 5137–5142.

CHAPTER 4

Antifungal Lipopeptides

4.1 INTRODUCTION

Microorganisms belonging to *Bacillus* spp are considered as biofactories for producing biologically active compounds that potentially inhibit growth of phytopathogens. Plant diseases have emerged as a global threat to food security. Fungi are major causative agents in plant diseases which lead to radical reductions in crop yield and cause economic loss to a country and its farmers. The antifungal lipopeptides are found to be involved in biocontrol activity against various phytopathogens such as fungi and bacteria that infect plants and crops (Ongena and Jacques 2008). Currently available antimicrobial agents used in agriculture are non biodegradable and are highly toxic and hence cause an increased risk of environment pollution. The compounds used as fungicides generate antifungal-resistant pathogens, which lead to potential risk of human health and environment (Northover and Zhou 2002). Hence, there is a need to develop novel, effective and safe compounds as antifungal drugs.

The *Bacillus* genus possesses a characteristic feature of producing a wide spectrum of secondary metabolites with antibiotic activity and diverse structures. Gene encoded antifungal agents are produced by different strains of *Bacillus subtilis* along with non-ribosomal synthesis of various small antibiotic peptides that are termed as lipopeptides. Lipopeptide antibiotics included in this category are surfactin (Kluge et al. 1988), fengycin (Vanittanakom et al. 1986) and iturin (Ongena and Jacques 2008, Gordillo and Maldonado 2012). The members of surfactin and fengycin consist of one hydroxyl fatty acid and seven and ten amino acids, respectively. Fengycin exhibits the primary function of inhibiting the growth of filamentous fungi (Vanittanakom, Loeffler et al. 1986). Surfactin possesses the ability to inhibit the growth of viruses, bacteria and cruors (Singh and Cameotra 2004). The mechanism of action of lipopeptides as an antifungal drug is an essential part to understand the development of novel bioactive compound. The fungitoxicity of fengycin and iturins depends upon their capability to permeate the cell membrane of the target microorganism (Deleu et al. 2003). The amphiphilic nature of antifungal lipopeptides such as fengycin, iturins and surfactin is because of the presence of a polar amino acid head and a hydrocarbon chain. They target the cell membrane and act by increasing the permeability of the cell leading to cell death (Maget-Dana et al. 1992, Deleu et al. 2008). *B. subtilis* fmbJ is capable of producing antifungal lipopeptides surfactin and fengycin (Huang et al. 2007). This is mainly a high fengycin producing strain among bacteria (Bie et al. 2009). All the families of antifungal *Bacillus* lipopeptides, fengycin, iturins and surfactins, have been explored by researchers for their proven antagonistic properties against a diverse category of phytopathogens which includes fungi, bacteria and oomycetes. Fengycin and iturins are known specifically for their antifungal activities while surfactin not only exhibits an antibacterial property but has also been explored as a larvicidal agent. The presence of surfactin along with other antifungal lipopeptides also enhances the effectivity of other lipopeptides (Maget-Dana et al. 1992, Tao et al. 2011). Furthermore, lipopeptides from the iturin family, being potent antifungal agents, are also used as biopesticides for protecting plants (Gordillo and Maldonado 2012). According to recent studies "Kannurin," a newly identified lipopeptide produced by *Bacillus cereus*, is also found to exhibit a potent antifungal property. These antifungal lipopeptides have been identified as potential weapons which can deal with an array of phytopathogens. The most interesting fact about lipopeptides is such molecules of biological origin are environmentally acceptable.

4.2 FENGYCIN

4.2.1 Introduction

An antifungal lipopeptide complex, fengycin, is produced by *B. subtilis* strain F-29-3. It exhibits inhibitory activity against filamentous fungi but cannot inhibit bacteria and yeast. This inhibition can be antagonized by phospholipids, oleic acids and sterols, whereas the other two unsaturated fatty acids participate in increasing the antifungal effect. The two main types of fengycin, fengycin A and fengycin B have a difference in only one amino acid. Fengycin A is composed of 1L-Ile, 1D-Ala, 1L-Pro,

1D-allo-Thr, 1D-Tyr, 3L-Glx, 1D-Orn, 1L-Tyr, while in fengycin B, D-Val is present instead of D-Ala. Both the analogs consist of variable lipid moiety, identified as anteiso-pentadecanoic acid (ai-C15), n-hexadecanoic acid (n-C16) and iso-hexadecanoic acid (i-C15), and further contain unsaturated and saturated residues up to C18. Originally the F-29-3 strain was described as a *Nocardia* species (Tschen and Liu 1977, Tschen and Liu 1978). Finally, it was identified as *B. subtilis* and recognized for its antifungal properties, specifically against *Rhizoctonia solani* through experiments performed in a laboratory and greenhouse with infected test plants (Tschen et al. 1982, Loeffler et al. 1986). The antifungal activity is also exerted by an antibiotic named bacilycin (bacilli-tetaine 5-10) (produced by *Bacillus amyloliquefaciens*) which possesses an inhibitory effect against budding fungi and bacteria. *Paecilomyces variotii*, *Rhizoctonia solani* and certain other filamentous fungi are insensitive to bacilycin. The inhibition of these organisms was carried out by other bioactive compound isolated by the same strain of bacteria (F-29-3). This antibiotic was formerly unknown, it is now described as fengycin. Its structure is composed of a β-hydroxy fatty acid linked to a peptide part of ten amino acids, where eight of them are organized in a cyclic structure (see Figure 4.1). This lipopeptide is known to possess antifungal activity against filamentous fungi and have hemolytic activity 40-fold lower than that of surfactin (1,3,4), another lipopeptide produced by *B. subtilis* (Deleu et al. 2008).

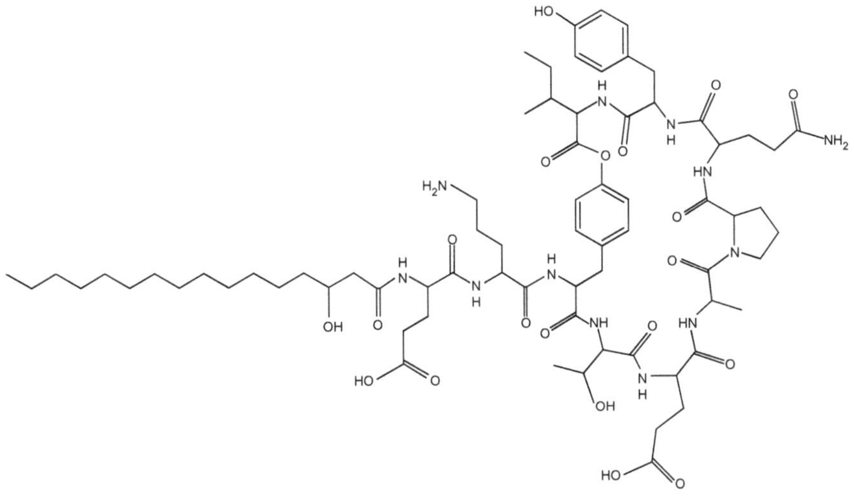

FIGURE 4.1   2D structure of fengycin.

4.2.2 Antifungal Properties of Fengycin

Similar to conventional antimicrobial peptides, fengycin also exhibits its activity via increasing the permeability of cell membrane of the target cell. The underlying molecular mechanism for membrane perturbation is not fully explained yet (Deleu et al. 2005). Because of the amphiphilic nature of fengycin with its hydrocarbon chain and a polar amino acid head it mainly targets cell membranes and increases cell permeability inducing cell death. The interaction of fengycin with lipid bilayer can be studied using membrane models such as monolayer model or lipid bilayer model. A recent study on a monolayer model showed that the membrane perturbing activity of fengycin on the morphology and structural characteristics of DPCC (dipalmitoyl-phosphatidyl-choline) monolayers is dependent on the concentration of fengycin. At low concentrations, fengycin is dispersed into the lipid bilayer as monomers and hence does not affect the phospholipid layer. At medium concentration, fengycin accumulates within the bilayer and self-associates, affecting the structure of the lipid bilayer. Higher concentrations of fengycin disrupt the lipid bilayer completely by forming mixed micelles (Deleu et al. 2008). Therefore, ion channel generation and pore formation is the most likely mechanism of fengycin for the disruption of the cell structure (Bechinger et al. 1993).

4.2.3 Identification of Ligands

The identification of ligands is an important step in recognizing specific sites of action for a drug molecule as unique ligands bind specifically with their corresponding drugs. For the identification of ligand, first three-dimensional (3D) structure of a drug needs to be recognized. The crystal structure of fengycin is not available in the PDB; hence homology modeling was performed for the prediction of the 3D structure of fengycin. BLAST result gave the maximum similarity with PDB ID 2VSQ (Surfactin A synthetase C SrfA-C, a non-ribosomal peptide synthetase termination module) for the amino acid sequence of fengyin. Further, homology modeling was performed to build the theoretical structure of fengycin. As DAB and 4FO are common ligands for lipopeptide, these were selected to see their interaction with the proposed model of fengycin (see Figure 4.2). As per the mechanism of action of fengycin described in the previous section, it is shown that the membrane perturbation activity of fengycin is due to the interaction with the membrane components such as sterols present on the surface of fungal cells. These sterols can be potential ligands. This is ensured by performing the molecular docking

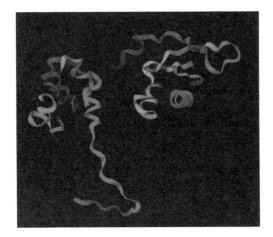

FIGURE 4.2   Structure of fengycin predicted through homology modeling.

of fengycin with various sterols i.e. ergosterol. Ligand identification is shown in Table 4.1.

4.2.4 Molecular Docking for Drug Targeting

Molecular docking for fengycin was performed using the glide application of Maestro suite of Schrödinger software. Docking studies were performed to elucidate the interaction of antifungal drug fengycin at its site of action on a fungal cell. In the previous step, ligands, i.e. DAB, 4FO and ergosterol

TABLE 4.1   Ligands of Fengycin with 2D Structures

Ligand	IUPAC Name & Molecular Formula	2D Structure
DAB	2,4-Diaminobutyric acid $C_4 H_{10} N_2 O_2$	
4FO	(2R)-2,4-diaminobutanoic acid $C_4 H_{10} N_2 O_2$	
ERG	Ergosterol	

were identified as potential sites on fungal cells for the action of fengycin. The process of molecular docking starts with the preparation of lipopeptide fengycin to optimize its structure. This step incorporates the removal of water molecules and restrains minimization of the molecule using a protein preparation wizard in the application tab of Maestro. This step generates the maximum possible orientations of protein structure. Further, using sitemap application, suitable sites are generated to perform docking. Ligands are docked against these generated sites to visualize this interaction in the binding pocket. Figure 4.3 represents interaction of fengycin with its potent target site ergosterol with the formation of four hydrogen bonds. The docking score and glide energy is −4.673 and −30.648 kcal/mol, respectively. In Figure 4.4, ligand 4FO is depicted in a binding pocket of fengycin. The interaction of fengycin and 4FO is established by the formation of four hydrogen bonds shown by dotted lines. The interaction has a docking score of −3.375 and glide energy is −20.559 kcal/mol. Similarly, Figure 4.5 depicts the interaction of ligand DAB and antifungal drug fengycin. The interaction is established through the formation of five hydrogen bonds and the docking score and glide energy is −3.405 and −20.475 kcal/mol, respectively. Table 4.2 summarizes the molecular docking results of fengycin with its ligands.

4.2.5 ADMET Studies of Fengycin

Simple prediction and integration of ADMET profiling (Absorption, distribution, metabolism, excretion and toxicity) of lead molecule is

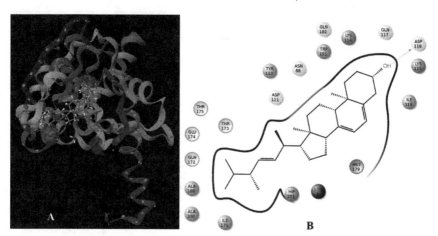

FIGURE 4.3   Hydrogen bond interaction (A) and ligand interaction diagram of fengycin with its potent target site ergosterol with the formation of four hydrogen bonds.

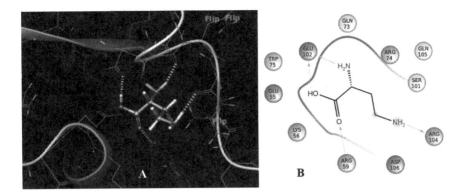

FIGURE 4.4   Hydrogen bond interaction (A) and ligand interaction diagram (B) between fengycin and 4FO.

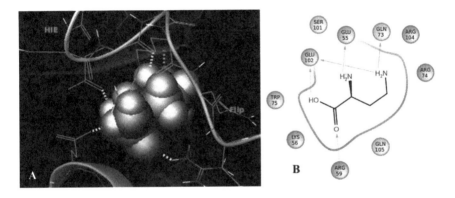

FIGURE 4.5   Hydrogen bond interaction (A) and ligand interaction diagram (B) between fengycin and DAB.

TABLE 4.2   Docking Results of Fengycin with its Ligands

Ligands	DAB	4FO	Ergosterol
H Bond	5	4	4
Docking Score	−3.405	−3.375	−4.673
Glide Energy (kcal/mol)	−20.475	−20.559	−30.648

performed to speed up the selection of lead for further clinical trials to save huge amount of cost for further experiments. ADMET prediction and profiling is typically dependent on various molecular descriptors such as Lipinski's "Rule of 5" (Ro5). Structural features of molecules are utilized by multiple approaches for the prediction of ADMET properties. Table 4.3 shows that the results for the blood-brain barrier (BBB) is negative (-ve),

TABLE 4.3   ADMET Study of Fengycin

ADMET Predicted Profile – Classification		
Model	Result	Probability
Absorption		
Blood-Brain Barrier	BBB-	0.9927
Human Intestinal Absorption	HIA+	0.5716
Caco-2 Permeability	Caco2-	0.8260
P-glycoprotein Substrate	Substrate	0.8419
P-glycoprotein Inhibitor	Non-inhibitor	0.8917
	Non-inhibitor	0.8705
Renal Organic Cation Transporter	Non-inhibitor	0.9215
Distribution		
Subcellular Localization	Lysosome	0.4960
Metabolism		
CYP450 2C9 Substrate	Non-substrate	0.8593
CYP450 2D6 Substrate	Non-substrate	0.8261
CYP450 3A4 Substrate	Substrate	0.5524
CYP450 1A2 Inhibitor	Non-inhibitor	0.9556
CYP450 2C9 Inhibitor	Non-inhibitor	0.9199
CYP450 2D6 Inhibitor	Non-inhibitor	0.8267
CYP450 2C19 Inhibitor	Non-inhibitor	0.8872
CYP450 3A4 Inhibitor	Non-inhibitor	0.7396
CYP Inhibitory Promiscuity	Low CYP Inhibitory Promiscuity	0.9318
Excretion		
Toxicity		
Human Ether-a-go-go-Related Gene Inhibition	Weak inhibitor	0.9340
	Non-inhibitor	0.5998
AMES Toxicity	Non-AMES toxic	0.7732
Carcinogens	Non-carcinogens	0.8368
Fish Toxicity	High FHMT	0.9823
Tetrahymena pyriformis Toxicity	High TPT	0.9883
Honey Bee Toxicity	Low HBT	0.7399
Biodegradation	Not readily biodegradable	0.9928
Acute Oral Toxicity	III	0.5685
Carcinogenicity (Three-class)	Non-required	0.5920
ADMET Predicted Profile – Regression		
Model	Value	Unit
Absorption		
Aqueous Solubility	−3.2679	LogS
Caco-2 Permeability	−0.2837	LogPapp, cm/s
		(*Continued*)

TABLE 4.3 (CONTINUED)   ADMET Study of Fengycin

	Distribution Metabolism Excretion Toxicity	
Rat Acute Toxicity	3.2233	LD50, mol/kg
Fish Toxicity	1.3798	pLC50, mg/L
Tetrahymena Pyriformis Toxicity	0.4597	pIGC50, ug/L

as negative value represents that an ideal drug should not cross the BBB. HIA+ shows that drug can be absorbed by human intestine and hence the most suitable method of drug intake is by oral administration. Other parameters show that fengycin is a non-inhibitor of other descriptors such as P-glycoprotein renal organic cation transporter and CYP450 enzymes. This depicts that fengycin does not interfere or inhibit the functions of these metabolic enzymes. ADMET results for toxicity and carcinogenicity prove the drug to be non-toxic and non-carcinogenic.

4.2.6 Pharmacophore Tool for Drug Discovery

Pharmacophore modeling is so far thriving and diverse section of computer-aided drug design. The pharmacophore concept is extensively practiced for rationale design of a novel drug. Computational execution of pharmacophore concept is very useful in the process of drug discovery. Pharmacophore models are used to depict and represent schematically two-dimensional and 3D elements of molecular recognition. The most common application of pharmacophore modeling is for target identification and virtual screening which deciphers possible strategies according to prior knowledge. Furthermore, pharmacophore concept is also helpful in off-target prediction, ADME-tox modeling as well as side effects. However, pharmacophore modeling is used in conjunction with molecular docking simulation to strengthen virtual screening. Here, pharmacophore mapping is performed to predict target proteins. It also includes studies on protein–protein interaction and inhibitor and protein design. (Figure 4.6).

4.3 ITURIN A

4.3.1 Introduction

Iturin is a class of lipopeptide antibiotics produced by *B. subtilis*. This class includes bacillomycin D, F, Lc, iturin A and mycosubtilin. All these are

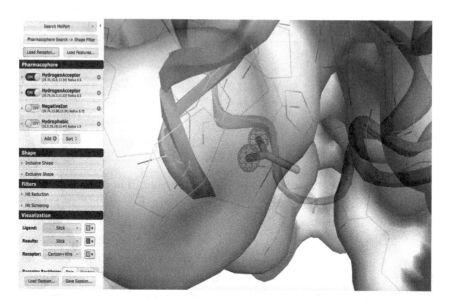

FIGURE 4.6 Pharmacophore mapping for fengycin.

cyclic lipopeptides and have β-amino fatty acids of C-14 to C-17 carbon atoms attached to a peptide of seven α-amino acids with configuration LDDLLDL (Aranda et al. 2005, Meena and Kanwar 2015). The first three amino acids in all three groups are common and all of them exhibit strong antifungal activity along with antibacterial and hemolytic properties. The amphiphilic structure of these lipopeptides indicates their strong interaction with cellular membrane constituents. Sterols are preferred target sites for their activity as sterols are mainly present on plasma membrane. The phenol ring on tyrosine residue of iturin compounds helps in the binding of these antibiotic compounds to plasma membrane. This results in distortion of cell membrane and lysis of cells.

B. subtilis produces three groups of lipopeptides: fengycin, iturin and surfactin groups. Fengycin and surfactin consist of one β-hydroxy fatty acid whereas the iturin class consists of β-amino fatty acid (Tsuge et al. 2001). Iturin A, the first antibiotic of the iturin family, was discovered in 1950 from the medium of *B. subtilis* (Nasir et al. 2013). The heptapeptide sequence of iturin A is L-Asn-D-Tyr-D-Asn-L-Gln-L-Pro-D-Asn-L-Ser. Iturin A forms ion-conducting pores in lipid bilayer. It disturbs the cellular membrane and modifies the phospholipid bilayer (Aranda et al. 2005). These molecules have varied applications in food and pharmaceutical industries due to their biological and physiochemical properties (Meena and Kanwar 2015).

Mycosubtilin is also produced by the *Bacillus* species and has anti-fungal activity. The peptide sequence of mycosubtilin consists of L-Asn-D-Tyr-D-Asn-L-Gln-L-Pro-D-Ser-L-Asn which was described in 1976 whereas bacillomycin F differs in the sixth and seventh amino acids (D-Asn-L-Thr instead of D-Ser-L-Asn). The iturin family of lipopeptides exhibit antifungal activities against *Penicillium chrysogenum*, *P. notatum*, *Absydia corymbifera*, *Trichophyton mentagrophytes*, *Aspergillus niger*, *Microsporum audouinii* and *Fusarium oxysporum*. They are also known to inhibit the growth of a few yeasts, namely, *Candida albicans*, *C. tropicalis* and *Saccharomyces cerevisiae* and two classes of bacteria (*Micrococcus* and *Sarcina*) (Nasir et al. 2013). Figure 4.7 represents the 2D structure of iturin which has been drawn through the 2D sketcher tool of Schrödinger Maestro software.

4.3.2 Mechanism of Action of Iturin A

Iturin forms ion-conducting pores in lipid membranes, which results in increased electrical conductance of the membranes. This leads to the leakage of potassium ions and other essential constituents of the cells leading to cell death (Maget-Dana et al. 1985, Maget-Dana and Peypoux 1994, Ongena and Jacques 2008). The primary mechanism of antibacterial activity of iturin A can be because of penetration of iturin A in the cytoplasmic membrane resulting in ion-conducting pores. Iturin A consists of an aliphatic hydrocarbon and a hydrophilic cyclic polypeptide. The hydrophobic moiety penetrates the hydrophobic areas of cell membrane whereas the hydrophilic part interacts with the polar groups of the membrane. It is still not clear if cytoplasmic membrane is the only site of action of iturin A.

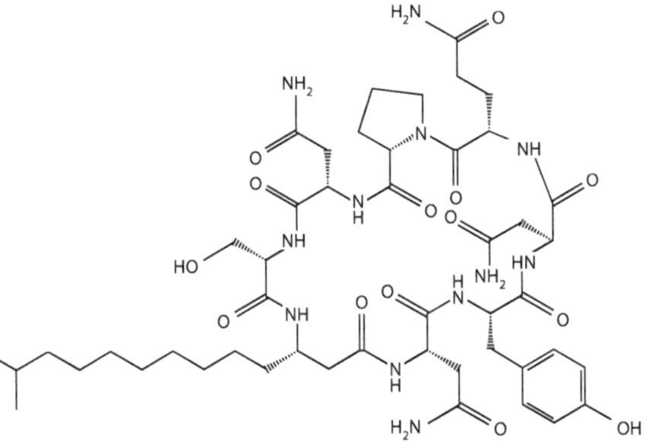

FIGURE 4.7   Structure of iturin.

In bacteria, the interaction of iturin A with cell membrane might lead to inactivation of metabolites or enzymes present on the cell membrane (Besson et al. 1978). Mycosubtilin is the most active compound of the iturin family. Mycosubtilin interacts with the acyl chains of phospholipids and alcohol group of sterols (particularly cholesterol) (Nasir and Besson 2011, Nasir et al. 2013, Gong et al. 2015).

4.3.3 Ligand of Iturin A

Antifungal lipopeptide iturin is experimentally reported in protein data bank (PDB) as pdb entry 2IHY. The rationalization of hypotheses generation and experimental data for new studies can be simulated through computational models. To model the mode of action of a drug, ligand docking at the binding site of a target receptor structure is performed. This further helps in pharmacological validation and optimization in medicinal chemistry. Pharmacophore elements are positioned according to the known ligands which generates the 3D representation of important features for activity and interaction of target. This also helps in the identification of new ligands of other chemical structures. The unique ligands are identified through the RCSB database and are represented in Table 4.4. D form of amino acids asparagine, tyrosine, serine and beta-alanine are such ligands are identified as DSG, DTY, DSN and BAL, respectively. These ligands are structural molecules of fungal cell membrane which can be available at the site of action of the antifungal drug iturin. The interaction of iturin and these ligands is validated using molecular docking simulation which can be justified through the positive interaction of drug and ligand.

4.3.4 Drug–Ligand Interaction by Molecular Docking

Molecular docking of iturin is performed in order to depict a computational model for the mechanism of action of iturin. The docking interaction depicts that iturin shows positive affinity toward certain amino acids of cell membrane protein through which iturin can bind to fungal cell membrane. This binding of amphiphilic natured iturin leads to pore formation on fungal cell membrane and ultimately disruption of cell membrane causing the fungal cell to die. The identified ligands in the previous step such as DSG, DTY, BAL and DSN are subjected to molecular docking against iturin using the Maestro suite of Schrödinger software. First the 3D structure of iturin is retrieved from PDB as pdb ID 2IHY. The protein structure is optimized through the protein preparation wizard.

Further, the docking site was generated through the sitemap and grid generation was performed through glide application. After grid generation each ligand was docked individually with a suitable docking site. Docking results are given in Table 4.5. Docking of ligand DSG, DTY, BAL and DSN with iturin (2IHY) was mediated by the formation of 6, 2, 1, 2 hydrogen bonds, respectively as depicted in Figures 4.8–4.11. The interaction has the

TABLE 4.4   Ligands of Iturin with 2D Structures

Ligand	IUPAC Name &Molecular Formula	2D Structure
DSG	D-Asparagine $C_4 H_8 N_2 O_3$	
DTY	D-Tyrosine $C_9 H_{11} N O_3$	
BAL	Beta-Alanine $C_3 H_7 N O_2$	
DSN	D-Serine $C_3 H_7 N O_3$	

TABLE 4.5   Docking Results of Iturin

Ligand	DSG	DTY	BAL	DSN
H Bond	6	2	1	2
Docking Score	−5.729	−4.773	−4.394	−3.178
Glide Energy (kcal/mol)	−27.268	−28.2720	−22.340	−22.498

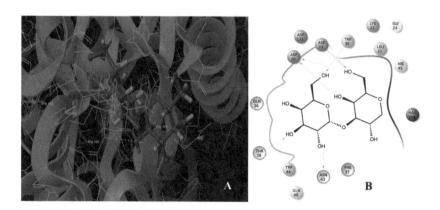

FIGURE 4.8 Hydrogen bond interaction (A) and ligand interaction diagram (B) between iturin and DSG.

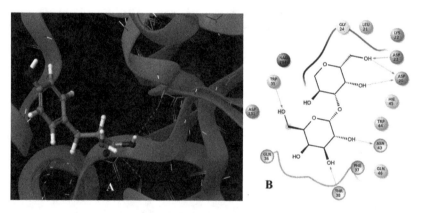

FIGURE 4.9 Hydrogen bond interaction (A) and ligand interaction diagram (B) between iturin and DTY.

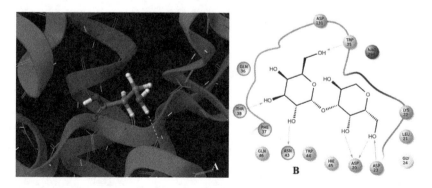

FIGURE 4.10 Hydrogen bond interaction (A) and ligand interaction diagram (B) between iturin and BAL.

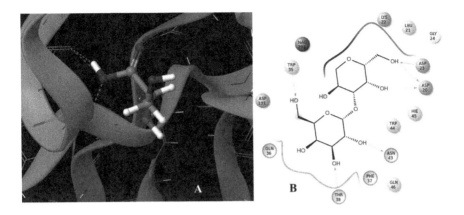

FIGURE 4.11   Hydrogen Bond Interaction (A) and Ligand Interaction Diagram (B) between Iturin and DSN.

docking score of −5.729, −4.773, −4.394, −3.178 for iturin and DSG, DTY, BAL, DSN, respectively. The glide energy for this interaction is −27.268, −28.2720, −22.340, −22.498 kcal/mol. According to the docking results, DSG, i.e. D-asparagine, has the highest negative docking score, which depicts the best interaction of iturin with ligand DSN with a minimum energy of −27.268 kcal/mol.

4.3.5   ADMET Modeling of Iturin A

The efficacy of a potential drug is dependent on the pharmacokinetic properties of the drug, also known as ADMET profiling, i.e. the characteristics of absorption, distribution, metabolism, excretion and toxicity. Generally, ADMET are rate limiting factors in the pipeline of drug discovery (DeLisle and Diller 2009). Computational analysis of ADMET profiling reduces the risk of a high attrition rate which could be expected in the later stages of drug design (Yamashita and Hashida 2004). ADMET profiling is also a promising approach for screening and optimization by testing only certain promising molecules. ADMET analysis in this work for antifungal drug, iturin, is performed through a web-based software admetSAR. This is a promising tool for predicting the various pharmacokinetic properties of drugs. Table 4.6 represents the ADMET analysis of iturin. According to the absorption profiling, BBB profiling predicted to be negative, which shows that iturin is not able to cross the BBB. Further human intestinal absorption is shown to be positive (HIA+). This suggests the drug is suitable for oral consumption. CaCO2 permeability is shown to be negative. The non-inhibitor for renal organic cation transporter significantly shows

TABLE 4.6   ADMET Profile of Iturin

ADMET Predicted Profile – Classification		
Model	**Result**	**Probability**
	Absorption	
Blood-Brain Barrier	BBB-	0.9867
Human Intestinal Absorption	HIA+	0.9188
Caco-2 Permeability	Caco2-	0.8252
P-glycoprotein Substrate	Substrate	0.8288
P-glycoprotein Inhibitor	Non-inhibitor	0.9458
	Non-inhibitor	0.9694
Renal Organic Cation Transporter	Non-inhibitor	0.8776
	Distribution	
Subcellular Localization	Mitochondria	0.4315
	Metabolism	
CYP450 2C9 Substrate	Non-substrate	0.8724
CYP450 2D6 Substrate	Non-substrate	0.7899
CYP450 3A4 Substrate	Non-substrate	0.5671
CYP450 1A2 Inhibitor	Non-inhibitor	0.9397
CYP450 2C9 Inhibitor	Non-inhibitor	0.9262
CYP450 2D6 Inhibitor	Non-inhibitor	0.8511
CYP450 2C19 Inhibitor	Non-inhibitor	0.8903
CYP450 3A4 Inhibitor	Non-inhibitor	0.9438
CYP Inhibitory Promiscuity	Low CYP Inhibitory Promiscuity	0.9742
	Excretion	
	Toxicity	
Human Ether-a-go-go-Related Gene Inhibition	Weak Inhibitor	0.9473
	Non-inhibitor	0.7693
AMES Toxicity	Non-AMES toxic	0.7613
Carcinogens	Non-carcinogens	0.8055
Fish Toxicity	High FHMT	0.9055
Tetrahymena pyriformis Toxicity	High TPT	0.9655
Honey Bee Toxicity	Low HBT	0.7768
Biodegradation	Not ready biodegradable	0.9792
Acute Oral Toxicity	III	0.6054
Carcinogenicity (Three-class)	Non-required	0.6480
ADMET Predicted Profile – Regression		
Model	**Value**	**Unit**
	Absorption	
Aqueous Solubility	−2.4929	LogS
Caco-2 Permeability	−0.0939	LogPapp, cm/s

(Continued)

TABLE 4.6 (CONTINUED) ADMET Profile of Iturin

	Distribution Metabolism Excretion Toxicity	
Rat Acute Toxicity	2.9684	LD50, mol/kg
Fish Toxicity	1.6972	pLC50, mg/L
Tetrahymena pyriformis Toxicity	0.3393	pIGC50, ug/L

the least risk of renal toxicity. Further distribution profile of drug is predicted in terms of its subcellular localization in mitochondria.

4.3.6 Pharmacophore Modeling of Iturin A

The pharmacophore should be considered as the largest common denominator of the molecular interaction features shared by a set of active molecules. Thus, a pharmacophore does not represent a real molecule or a set of chemical groups, but is an abstract concept. Despite this clear definition, the term pharmacophore is often misused by many in medicinal chemistry to describe simple yet essential chemical functionalities in a molecule. Often the long definition is simplified to "A pharmacophore is the pattern of features of a molecule that is responsible for a biological effect," which captures the essential notion that a pharmacophore is built from features rather than defined chemical groups. A pharmacophore is the ensemble of steric and electronic features that is necessary to ensure the optimal supramolecular interactions with a specific biological target and to trigger (or block) its biological response. Pharmacophore modeling for Iturin lipopeptide is performed using the reported NMR (Nuclear Magnetic Resonance) structure on the RCSB website with PDB entry 2IH0 (see Figure 4.12).

4.4 SURFACTIN

4.4.1 Introduction

Surfactin, a cyclic lipopeptide produced by bacteria, is a powerful surfactant and generally used as an antibiotic (Mor 2000). This substance is capable of surviving in both hydrophobic and hydrophilic environments because of its amphiphilic nature. Surfactin antibiotic is produced by *B. subtilis*, a Gram-positive endospore forming bacteria (Peypoux et al. 1999). Various studies pertaining to properties of surfactin reveal that it exhibits the potent characteristics of being antifungal, antiviral, antibacterial, hemolytic and having antimycoplasma activities (Singh and

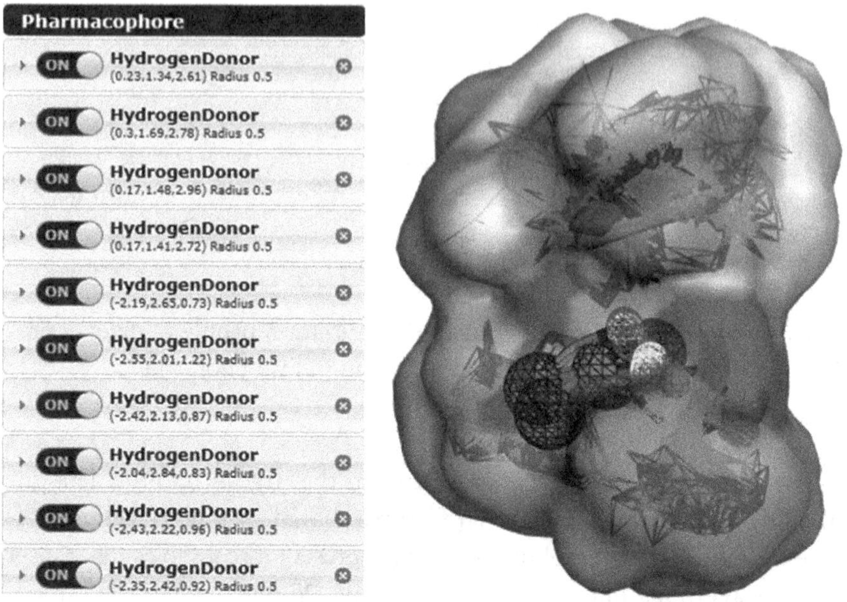

FIGURE 4.12   Pharmacophore mapping for fengycin (2IH0 with BAL).

Cameotra 2004). The structure of surfactin consists of a peptide loop of seven amino acids (L-leucine, L-aspartic acid, L-valine, glutamic acid and two D-leucines) and a thirteen to fifteen carbon long hydrophobic fatty acid chain which facilitates its penetration into cellular membranes (see Figure 4.13). Glutamic acid at position 1 and aspartic acid at position 5 constitute a minor polar domain while fatty acid makes up a major hydrophobic domain on valine at 4th position. The fatty acid tail can extend freely into solution below the critical micellar concentration (CMC) and further participate in hydrophobic interactions within micelles (Grau et al. 1999). Various studies pertaining to properties of surfactin reveal that it exhibits the potent characteristics of being antifungal, antiviral, antibacterial, hemolytic and having antimycoplasma activities (Singh and Cameotra 2004).

Surfactin is synthesized through a linear non-ribosomal peptide synthetase by surfactin synthetase. In solution, it exerts "horse saddle" like conformation which explains its wide range of biological activity (Hue et al. 2001). Surfactin changes the surface tension of the liquid it is dissolved in. It possesses the ability to lower surface tension of water from 72 mN/m to 27 mN/m with their minimum concentration of 20 μM (Yeh et al. 2005) which is because it occupies the intermolecular spaces between

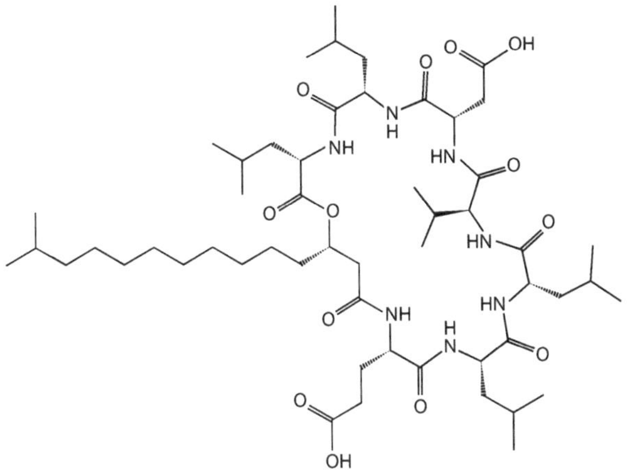

FIGURE 4.13   Chemical structure of surfactin.

water molecules. Surfactin derived from *Bacillus circulans* is also active against multidrug-resistant bacteria like *Proteus vulgaris, Alcaligenes faecalis, Escherichia coli, Pseudomonas aeruginosa,* and methicillin-resistant *Staphylococcus aureus* (Das et al. 2008). The minimal bactericidal and minimal inhibitory concentrations of surfactin is also comparatively lower than some of the conventional antibiotics (Das et al. 2008). Mycoplasma is a parasite of eukaryotic cells, the smallest free-living organism and one of the foremost contaminants affecting culture cells of mammalian tissue. Mycoplasma are severe causative agents for both animal and human diseases, which include urogenital tract infections, acute respiratory inflammations (including pneumonia) and AIDS (Organization 1993, Blanchard and Montagnier 1994). Treatment incorporating the use of antibiotics is the most efficient process for suppression and elimination of the infection of mycoplasma in cell cultures. Surfactin is commercially used to treat cell cultures and to get rid of mycoplasmic contamination in biotechnological products (Boettcher et al. 2010).

4.4.2 Mode of Action – Surfactin

Surfactin is a versatile bioactive compound with antimycoplasma activity (Sen 2010). The disintegration of mycoplasma is apparently because of the physicochemical interaction of the outer part of the lipid bilayer membrane with the membrane-active surfactin. This leads to alteration in permeability of the cell membrane lipids which cause leakage of ions from the cell into the surrounding tissue. Higher concentration of surfactin results in the disintegration of the membrane system of mycoplasma due to its detergent-like property (Meena and Kanwar 2015). The antibiotic nature of surfactin makes it suitable as an antibacterial agent too, as it exerts the ability to penetrate the cell membrane of bacteria of all types. Surfactin also exhibits an antiviral property as it can disintegrate viral envelopes and capsid by forming ion channels.

4.4.3 Discovering Ligands of Surfactin

Ligand discovery for the interpretation of the key biological process of ligand-drug interaction is performed at molecular level. This includes the characterization and identification of small molecules' binding sites on proteins of therapeutic interest. It has remarkable implications for target identification in the drug discovery process (Guo et al. 2015). Ligand identification for surfactin is done on the basis of template protein used to predict the 3D structure of surfactin. The 3D structure of surfactin is predicted by homology modeling. The generation of model was done on template protein with pdb ID 2RON (Structure of Chain A of surfactin synthetase thioesterase module). 2RON was selected on the basis of sequence alignment through BLAST which gave 76.34% similarity with a surfactin sequence of *B. subtilis*. The structure of surfactin as a result of homology modeling is depicted in Figure 4.14. Ligand identification was done using the RCSB database of proteins. Leucine is identified as a potential ligand for the surfactin thioesterase domain. This is further justified with molecular docking analysis which depicts the potential interaction of surfactin with the ligand leucine (see Table 4.7).

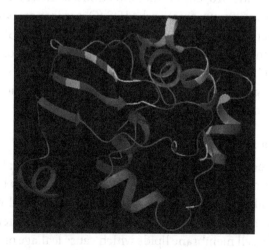

FIGURE 4.14   3D view of surfactin by homology modeling.

TABLE 4.7   Ligands of Surfactin with 2D Structures

Ligand	IUPAC Name &Molecular Formula	2D Structure
LEU	Leucine $C_6H_{13}NO_2$	

4.4.4 Molecular Docking as a Tool for Design of Drugs

Molecular docking analysis is done in order to depict the mode of action of surfactin, where the interaction of surfactin is depicted with the ligand leucine. The amino acid leucine, being a part of the surface protein of microorganisms, provides the suitable site of interaction for surfactin as a drug to act upon. To perform docking, the glide module of Schrödinger maestro suite is used. It was started with the protein preparation wizard for restraining minimization and optimization of the surfactin 3D structure. A sitemap was used to generate a suitable site in the lipopeptide surfactin. This resulted in various suitable orientations of protein and its binding sites. Subsequently, the 2D structure of leucine was drawn using a 2D sketcher tool. Further glide application was used to perform molecular docking of surfactin and leucin and binding sites from project table entries. As per Figure 4.15 the interaction is established through the formation of four hydrogen bonds depicted in Figure B. Figure 4.15 depicts the ligand interaction diagram of leucine surrounded with amino acids in a binding pocket of sufactin. This shows that ASN 109 (asparagine at position 109) is participating in the formation of three hydrogen bonds with an amino and hydroxyl group of leucine. Similarly, TYR at 67 position is interacting with carbonyl oxygen. This

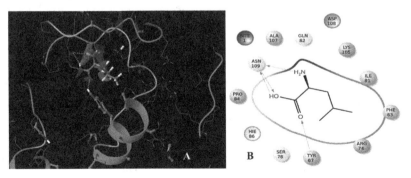

FIGURE 4.15   Molecular docking of surfactin (A) hydrogen bond interaction diagram (B) ligand interaction diagram for surfactin and leucine.

interaction has resulted in the docking score of −4.010 and glide energy of −22.904 kcal/mol (see Table 4.8).

4.4.5 ADMET Studies of Surfactin

The ADMET properties of surfactin have been predicted through the online web-based software named admetSAR. The tool has been a reliable source of predicting druglikeness and pharmacokinetic properties of a drug candidate. According to the report generated (see Table 4.9), the absorption profiling shows that because of the high molecular weight of the drug surfactin, it is not able to cross the BBB. For the same reason of high molecular weight, low intestinal absorption of the drug is also shown. This parameter can be helpful in treating a tumor as the drug would be more effective if the tumor is directly and specifically targeted. In certain cases, absorption also leads to low availability of drug at target sites which can be suitable to treating infection of the GI (Gastrointestinal) tract as an antibacterial drug because surfactin is reported to possess antibacterial and antiviral properties. BBB −ve profiling of a drug is considered to check its effect on the brain. Hence, ideally drugs do not cross the BBB. Further, Caco2 permeability is also reported to be negative. Surfactin is a non-inhibitor of P-glycoprotein and renal organic cation transporter, this suggests the non-interference of drug with general metabolic functions of the body. Subcellular localization of the drug is distributed in mitochondria. The metabolism profiling predicts that surfactin is a non-substrate and a non-inhibitor of CYP450 (Cytochrome P450) enzymes. Toxicity profiling depicts the drug to be non-AMES toxic, non-carcinogenic, with low honey bee toxicity and high FHMT (Fathead Minnow Toxicity), as well as high TPT (*Tetrahymena pyriformis* toxicity).

The drawback of surfactin as a drug is its nonspecific toxicity. It has the ability to kill normal cells apart from the pathogenic cells. The haemolytic property is also resulting of its ability to lyse red blood cells (RBCs). Because of this, caution must be taken if the drug is administered intravascularly. This nonspecific cytotoxicity is fortunately at a higher concentration which is approximately 40 μM to 60 μM. This is also the LD50 value for proliferating cells *in-vitro*. The toxicity effect of surfactin is not significant at concentrations below 25 μM.

TABLE 4.8   Docking Results of Surfactin

Ligand	LEU
H Bond	4
Docking Score	−4.010
Glide Energy (kcal/mol)	−22.904

TABLE 4.9   ADMET Profile of Surfactin

ADMET Predicted Profile – Classification		
Model	**Result**	**Probability**
	Absorption	
Blood-Brain Barrier	BBB-	0.9427
Human Intestinal Absorption	HIA+	0.8150
Caco-2 Permeability	Caco2-	0.7683
P-glycoprotein Substrate	Substrate	0.7118
P-glycoprotein Inhibitor	Non-inhibitor	0.6335
	Non-inhibitor	0.7749
Renal Organic Cation Transporter	Non-inhibitor	0.9635
	Distribution	
Subcellular Localization	Mitochondria	0.6768
	Metabolism	
CYP450 2C9 Substrate	Non-substrate	0.8585
CYP450 2D6 Substrate	Non-substrate	0.8413
CYP450 3A4 Substrate	Non-substrate	0.5311
CYP450 1A2 Inhibitor	Non-inhibitor	0.9491
CYP450 2C9 Inhibitor	Non-inhibitor	0.9410
CYP450 2D6 Inhibitor	Non-inhibitor	0.9416
CYP450 2C19 Inhibitor	Non-inhibitor	0.9381
CYP450 3A4 Inhibitor	Non-inhibitor	0.7473
CYP Inhibitory Promiscuity	Low CYP Inhibitory Promiscuity	0.9847
	Excretion	
	Toxicity	
Human Ether-a-go-go-Related Gene Inhibition	Weak Inhibitor	0.9816
	Non-inhibitor	0.9597
AMES Toxicity	Non-AMES toxic	0.8964
Carcinogens	Non-carcinogens	0.9542
Fish Toxicity	High FHMT	0.6914
Tetrahymena pyriformis Toxicity	High TPT	0.9862
Honey Bee Toxicity	Low HBT	0.7184
Biodegradation	Not ready biodegradable	0.9443
Acute Oral Toxicity	III	0.6639
Carcinogenicity (Three-class)	Non-required	0.6658
ADMET Predicted Profile – Regression		
Model	**Value**	**Unit**
	Absorption	
Aqueous Solubility	−2.2693	LogS
Caco-2 Permeability	−0.1510	LogPapp, cm/s

(*Continued*)

TABLE 4.9 (CONTINUED)   ADMET Profile of Surfactin

	Distribution Metabolism Excretion Toxicity	
Rat Acute Toxicity	2.8726	LD50, mol/kg
Fish Toxicity	1.9929	pLC50, mg/L
Tetrahymena pyriformis Toxicity	0.2581	pIGC50, ug/L

4.4.6 Pharmacophore Studies in Drug Design

The pharmacophore concept shows the molecular orientation from three dimensions. It has nevertheless become an important tool in CADD. Every atom or group in a molecule that exhibits certain properties related to molecular recognition can be reduced to a pharmacophore feature. These molecular patterns can be labeled as hydrogen bond donors or acceptors, cationic, anionic, aromatic, or hydrophobic, and any possible combinations. Different molecules can be compared at the pharmacophore level; this usage is often described as "pharmacophore fingerprints." When only a few pharmacophore features are considered in a 3D model the pharmacophore is sometimes described as a "query." The structure used for pharmacophore modeling of surfactin is used from RCSB with PDB ID 2NPV which is an NMR studied structure and dynamics of surfactin in micellar media (see Figures 4.16 and 4.17).

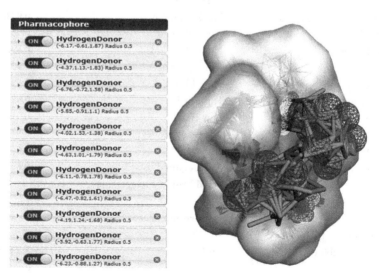

FIGURE 4.16   Pharmacophore mapping of surfactin 2NPV.

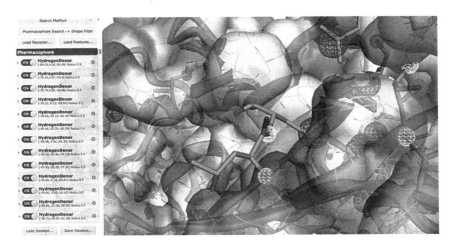

FIGURE 4.17   Zoomed view of pharmacophore model of surfactin.

4.5 SUMMARY

Fungi are one of the major reasons for plant diseases resulting in reduced crop yield. The conventional chemical fungicides cause strain resistance among fungal species and hence there is a need to develop new and safe antifungal drugs with novel targets. Ergosterol is the major component of lipid membranes, hence it is a suitable target for any drug to act on it. Fengycin generally targets DPCC (dipalmitoyl-phosphatidyl-choline) monolayers present on the membranes which suggest its potential to bind to ergosterols. As depicted by molecular docking studies, fengycin interacts with ergosterol via four hydrogen bonds resulting in a docking score of −4.673 and glide energy of −30.648 kcal/mol which is significantly higher than the remaining two ligands of fengycin. On the other hand, the interaction of iturin and surfactin is only with few amino acids. These amino acids may or may not be present as the membrane proteins and hence they cannot be considered as a suitable docking site for the compounds to act upon. This shows that fengycin is the best drug among the three molecules discussed in this chapter.

REFERENCES

Aranda, F. J., Teruel, J. A., Ortiz, A. 2005. Further aspects on the hemolytic activity of the antibiotic lipopeptide iturin A. *Biochimica et Biophysica Acta (BBA)-Biomembranes* 1713(1): 51–56.

Bechinger, B., Zasloff, M., Opella, S. J. 1993. Structure and orientation of the antibiotic peptide magainin in membranes by solid-state nuclear magnetic resonance spectroscopy. *Protein Science* 2(12): 2077–2084.

Besson, F., Peypoux, F., Michel, G., Delcambe, L. 1978. Mode of action of itu-rin A, an antibiotic isolated from *Bacillus subtilis*, on Micrococcus luteus. *Biochemical and Biophysical Research Communications* 81(2): 297–304.

Bie, X., Lu, Z., Lu, F. 2009. Identification of fengycin homologues from *Bacillus subtilis* with ESI-MS/CID. *Journal of Microbiological Methods* 79(3): 272–278.

Blanchard, A. and Montagnier, L. 1994. AIDS-associated mycoplasmas. *Annual Review of Microbiology* 48(1): 687–712.

Boettcher, C., Kell, H., Holzwarth, J. F., Vater, J. 2010. Flexible loops of thread-like micelles are formed upon interaction of L-α-dimyristoyl-phosphatidylcholine with the biosurfactant surfactin as revealed by cryo-electron tomography. *Biophysical Chemistry* 149(1–2): 22–27.

Das, P., et al. 2008. Antimicrobial potential of a lipopeptide biosurfactant derived from a marine *Bacillus circulans*. *Journal of Applied Microbiology* 104(6): 1675–1684.

Deleu, M., Bouffioux, O., Razafindralambo, H., et al. 2003. Interaction of surfac-tin with membranes: a computational approach. *Langmuir* 19(8): 3377–3385.

Deleu, M., Paquot, M., Nylander, T. 2005. Fengycin interaction with lipid mono-layers at the air–aqueous interface—implications for the effect of fengycin on biological membranes. *Journal of Colloid and Interface Science* 283(2): 358–365.

Deleu, M., Paquot, M., Nylander, T. 2008. Effect of fengycin, a lipopeptide pro-duced by *Bacillus subtilis*, on model biomembranes. *Biophysical Journal* 94(7): 2667–2679.

DeLisle, R. K. and Diller, D. J. 2009. Edtiorial [Hot topic: in silico ADME/Tox models: progress and challenges (Guest editors: Robert K. DeLisle and David J. Diller)]. *Current Computer-Aided Drug Design* 5(2): 69–70.

Dufour, S., Deleu, M., Nott, K., et al. 2005. Hemolytic activity of new linear sur-factin analogs in relation to their physico-chemical properties. *Biochimica et Biophysica Acta (BBA)-General Subjects* 1726(1): 87–95.

Gong, A. D., Li, H. P., Yuan, Q. S., et al. 2015. Antagonistic mechanism of iturin A and plipastatin A from *Bacillus amyloliquefaciens* S76-3 from wheat spikes against *Fusarium graminearum*. *PloS One* 10(2): e0116871.

Gordillo, A. and Maldonado, M. C. 2012. Purification of peptides from *Bacillus* strains with biological activity. *Chromatography and Its Applications* 11: 201–225.

Grau, A., Fernández, J. C. G., Peypoux, F., Ortiz, A. 1999. A study on the interac-tions of surfactin with phospholipid vesicles. *Biochimica et Biophysica Acta (BBA)-Biomembranes* 1418(2): 307–319.

Guo, Z., Li, B., Cheng, L.-T., et al. 2015. Identification of protein–ligand bind-ing sites by the level-set variational implicit-solvent approach. *Journal of Chemical Theory and Computation* 11: 753–765.

Huang, X., Lu, Z., Bie, X., et al. 2007. Optimization of inactivation of endospores of *Bacillus cereus* by antimicrobial lipopeptides from *Bacillus subtilis* fmbj strains using a response surface method. *Applied Microbiology and Biotechnology* 74(2): 454–461.

Hue, N., Serani, L., Laprévote, O. 2001. Structural investigation of cyclic pepti-dolipids from *Bacillus subtilis* by high-energy tandem mass spectrometry. *Rapid Communications in Mass Spectrometry* 15(3): 203–209.

Kluge, B., Vater, J., Salnikow, J., Eckart, K. 1988. Studies on the biosynthesis of surfactin, a lipopeptide antibiotic from *Bacillus subtilis* ATCC 21332. *FEBS Letters* 231(1): 107–110.

Loeffler, W., Tschen, J. S. M., Vanittanakom, N., et al. 1986. Antifungal effects of bacilysin and fengymycin from *Bacillus subtilis* F-29-3 a comparison with activities of other *Bacillus* antibiotics. *Journal of Phytopathology* 115: 204–213.

Maget-Dana, R. and Peypoux, F. 1994. Iturins, a special class of pore-forming lipopeptides: biological and physicochemical properties. *Toxicology* 87(1–3): 151–174.

Maget-Dana, R., Ptak, M., Peypoux, F., Michel, G. 1985. Pore-forming properties of iturin A, a lipopeptide antibiotic. *Biochimica et Biophysica Acta (BBA)-Biomembranes* 815(3): 405–409.

Maget-Dana, R., Thimon, L., Peypoux, F., Ptak, M. 1992. Surfactin/iturin A interactions may explain the synergistic effect of surfactin on the biological properties of iturin A. *Biochimie* 74(12): 1047–1051.

Meena, K. R. and Kanwar, S. S. 2015. Lipopeptides as the antifungal and antibacterial agents: applications in food safety and therapeutics. *BioMed Research International* 2015, Article ID 473050, 9 pages.

Mor, A. 2000. Peptide-based antibiotics: A potential answer to raging antimicrobial resistance. *Drug Development Research* 50(3–4): 440–447.

Nasir, M. N. and Besson, F. 2011. Specific interactions of mycosubtilin with cholesterol-containing artificial membranes. *Langmuir* 27(17): 10785–10792.

Nasir, M. N., Besson, F., Deleu, M. 2013. Interactions of iturinic antibiotics with plasma membrane. Contribution of biomimetic membranes. *Biotechnologie, Agronomie, Société et Environnement* 17(3): 505–516.

Northover, J. and Zhou, T. 2002. Control of rhizopus rot of peaches with postharvest treatments of tebuconazole, fludioxonil, and *Pseudomonas syringae*. *Canadian Journal of Plant Pathology* 24(2): 144–153.

Ongena, M. and Jacques, P. 2008. Bacillus lipopeptides: versatile weapons for plant disease biocontrol. *Trends in Microbiology* 16(3): 115–125.

Organization, W. H. 1993. Report of the WHO Meeting on the Development of Vaginal Microbicides for the Prevention of Heterosexual Transmission of HIV. World Health Organization, Geneva, Switzerland.

Peypoux, F., Bonmatin, J. M., Wallach, J. 1999. Recent trends in the biochemistry of surfactin. *Applied Microbiology and Biotechnology* 51(5): 553–563.

Sen, R. 2010. Surfactin: Biosynthesis, Genetics and Potential Applications. In: Sen R. (eds) Biosurfactants. Advances in Experimental Medicine and Biology, vol 672. Springer, New York, NY.

Singh, P. and Cameotra, S. S. 2004. Potential applications of microbial surfactants in biomedical sciences. *Trends in Biotechnology* 22(3): 142–146.

Tao, Y., Bie, X. M., Lv, F. X., Zhao, H. Z., Lu, Z. X. 2011. Antifungal activity and mechanism of fengycin in the presence and absence of commercial surfactin against *Rhizopus stolonifer. The Journal of Microbiology* 49(1): 146–150.

Toraya, T., Maoka, T., Tsuji, H., Kobayashi, M. 1995. Purification and structural determination of an inhibitor of starfish oocyte maturation from a *Bacillus* species. *Applied and Environmental Microbiology* 61(5): 1799–1804.

Tschen, J., Kuo, W. and Liu, J. 1982. Inhibition of Rhizoctonia solani by *Bacillus* antagonist. *Li kung hsueh pao= Journal of Science &Engineering.*

Tschen, J. and Liu, J. 1977. *Nocardia* sp. as antagonists to *Rhizoctonia solani. Plan Protection Bulletin (Taipei)* 19: 301.

Tschen, J. and Liu, J. 1978. Antagonistic effects of *Nocardia* sp. on *Rhizoctonia solani. Third International Congress of Plant Pathology 3rd Internat. Congr. Plant Pathol.*, Munchen.

Tsuge, K., Akiyama, T., Shoda, M. 2001. Cloning, sequencing, and characterization of the iturin A operon. *Journal of Bacteriology* 183(21): 6265–6273.

Vanittanakom, N., Loeffler, W., Koch, U., Jung, G. 1986. Fengycin-a novel antifungal lipopeptide antibiotic produced by *Bacillus subtilis* F-29-3. *The Journal of Antibiotics* 39(7): 888–901.

Yamashita, F. and Hashida, M. 2004. In silico approaches for predicting ADME properties of drugs. *Drug Metabolism and Pharmacokinetics* 19(5): 327–338.

Yeh, M. S., Wei, Y. H., Chang, J. S. 2005. Enhanced Production of Surfactin from *Bacillus subtilis* by addition of solid carriers. *Biotechnology Progress* 21(4): 1329–1334.

CHAPTER 5

Precursors of Lipopeptides

5.1 PLIPASTATIN SYNTHASE

5.1.1 Introduction

Non-ribosomal peptide synthetases (NRPSs) are enzymatic complexes which act as catalyzers for the synthesis of biologically important microbial peptides. One of these microbial peptides of commercial importance is cyclic lipopeptides. Cyclic lipopeptides (CLPs) are structurally and functionally diverse molecules, which are produced by large, complex and multifunctional NRPSs using the thioesterase mechanism. NRPSs acquire a modular structure and each module acts as a building block resulting in a stepwise incorporation and modification of one amino acid unit (Raaijmakers et al. 2006). This shows their highly conserved structural organization into specific peptide binding domain (Tosato et al. 1997). Plipastatin is one such example of a cyclic lipodecapeptide. An antifungal lipopeptide, plipastatin, was initially isolated as an inhibitor of phospholipase from *Bacillus cereus* strain BMG302-fF67.

Plipastatin is composed of a ten-amino-acid cyclic peptide and an attached β-hydroxy fatty acid chain of 14 to 18 carbon atoms at N-terminus of peptide (see Figure 5.1). Plipastatin has an assembly of five large multienzyme NRPSs, namely, PPSA, PPSB, PPSC, PPSD and PPSE and are encoded by genes *ppsA*, *ppsB*, *ppsC*, *ppsD* and *ppsE* in the operon of plipastatin synthetase, respectively. The NRPSs are subdivided into various modules which are responsible for the initiation, elongation

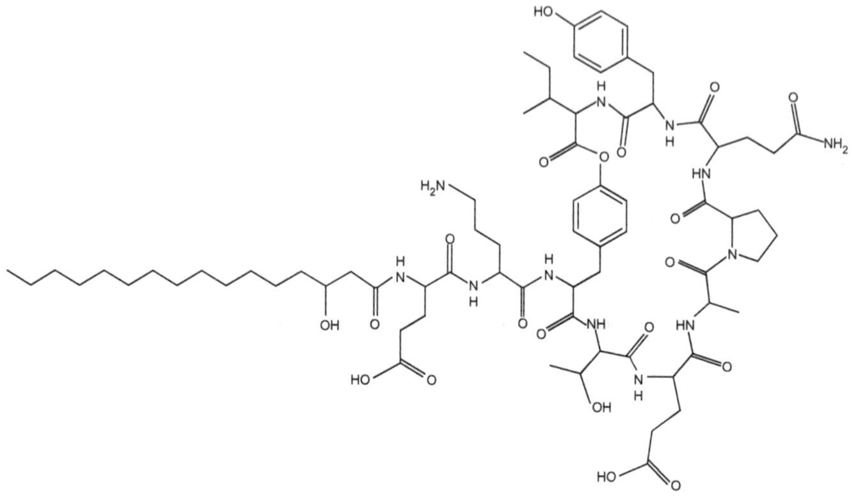

FIGURE 5.1   2D structure of plipastatin.

and termination according to their biosynthetic functions. Each module consists of three domains such as the condensation domain (C) which is responsible for the formation of peptide bonds. The adenylation domain (A) is responsible for the recognition and activation of amino acids. Finally, the thiolation domain (T), which is also known as peptidyl carrier protein (PCP), is responsible for tethering the activated substrates and elongating the chain of peptides. Certain modules also contain an epimerase domain that is involved in converting the L form of amino acids into its D isomers. The termination module often has a thioesterase (TE) domain in NRPSs that is responsible for cyclization and release of product.

Plipastatin was initially reported in 1986 by Umezwa et al. (1986). As per the nomenclature, plipastatin is identical to fengycin with some structural variations at various saline conditions. The difference between fengycin and plipastatin is in the orientation of Tyr3 and Tyr9, respectively in L and D form. These are produced by various strains of *Bacillus* species and are encoded by gene *Pps* for various modules of plipastatin synthase. A large and multifunctional enzyme complex, plipastatin synthase is involved in the activation and polymerization of the amino acids proline, glycine and tyrosine as part of plipastatin synthesis, which is a lipopeptide antibiotic (Tsuge et al. 1999). Subunit D of plipastatin synthase (ppsD) is composed of five NRPS subunits. These assemble and compile to form a linear chain of order ppsC-ppsD-ppsE-ppsA-ppsB. The structure is comprised of a β-hydroxy fatty acid chain connected to a ten-amino-acid long peptide part in which eight amino acids participate in forming a cyclic structure. There is loss of the production of plipastatin if the ppsD domain is disrupted (Steller et al. 1999, Lin et al. 2005, Ongena et al. 2005).

5.1.2 Mechanism of Action of the Corresponding Lipopeptide

Evaluation of the mechanism of action of plipastatin from the *Bacillus* species was performed against plant pathogen *Fusarium graminearum*. Plipastatin has shown discrete antagonistic activity against fungal plant pathogen and hence is categorized as antifungal lipopeptide. Microscopic and time-lapse imaging analyses revealed that plipastatin A mainly causes vacuolation and conglobation on young hyphae and branch tips. Transmission electron microscopy (TEM) showed that treatment of *F. graminearum* with compound creates cell walls with wide gaps and disturbs plasma membranes. The results of these analyses indicate that plipastatin A has antagonistic mechanisms against *F. graminearum* and leads to deleterious cellular consequences in *F. graminearum*. *Bacillus* strain S76-3 that can produce large quantities of plipastatin compound has significant potential for its use as a biocontrol agent for controlling *Fusarium* pathogens in agriculture (Gong et al. 2015). Plipastatin, being an antifungal lipopeptide, is structurally related to fengycin and the iturin family of antifungal lipopeptides produced by various species of *Bacillus*. Fengycin, reported by Budzikiewicz et al. in 1999, and plipastatin, reported by Umezawa et al. in 1986, have marginal difference in their structures (Honma et al. 2012). Fengycin has already been shown to possess various bioactivities leading to inhibition of fungal growth (Vanittanakom et al. 1986). From the structural similarity of plipastatin with fengycin, it can be stated that these antifungal lipopeptides cause bending, membrane perturbation and micelle formation in experiments on artificial membranes (Deleu et al. 2005, Patel et al. 2011, Horn et al. 2013).

The antifungal activity of fengycin is considered to be associated with sterols and phospholipids in cell membranes. Fengycin acts by forming pores on cell membrane via an all or none mechanism wherein fengycin at low concentration shows no effect on cell membrane while high concentration of fengycin causes large pores on affected cells allowing thorough leakage of intracellular components with death of the cell (Falardeau et al. 2013). According to the studies on antifungal lipopeptide iturin, the electrical conductance increases by formation of ion-conducting pores on artificial lipid membranes in yeast cells (Maget-Dana et al. 1985). This also leads to disturbance in cytoplasmic membrane and causes leakage of potassium ion (K⁺) and other vital components of the cell resulting in cell death (Maget-Dana and Peypoux 1994, Ongena and Jacques 2008). The studies of lipopeptides on artificial membranes showed that the action of lipopeptides depends upon the interaction of various sterol components of cell membrane and phospholipids (Nasir et al. 2010). This interaction is mediated by an alcohol group of cholesterol and acyl chain of phospholipids in membrane. The preferential affinity of lipopeptides is reported specifically for cholesterol in animal cells as compared to the ergosterol which is a main fungal sterol (Gong et al. 2015).

5.1.3 Ligand Identification of Plipastatin Synthase

Ligand identification of plipastatin is done on the basis of its mechanism of action and its structure prediction through homology modeling on the NRPS template. Plipastatin is described as an antifungal drug. Its mechanism of action states that ergosterol, the main fungal sterol, is a potential site of action to destroy cell membrane of fungal cells. Lipopeptides, by nature, possess affinity toward the sterol components of cell membrane. To validate the antifungal activity of plipastatin, ergosterol is chosen as one of its ligands. Another ligand chosen is leucine, which is a common ligand for NRPS template protein with PDB ID 2VSQ. Table 5.1 summarizes the chemical formula and structure of the ligands of plipastatin.

5.1.4 Structure Determination of Plipastatin Synthase Using Homology Modeling

The structure of plipastatin synthase was generated by homology modeling. The lipopeptides consist of polypeptide in their composed structure. It is known that three-dimensional structures of proteins are more conserved than their primary sequence; hence the structural analysis has

TABLE 5.1   Ligands for Plipapstatin

Ligand	IUPAC Name & Molecular Formula	2D Structure
Leu	Leucine $C_6H_{13}NO_2$	
ERG	Ergosterol $C_{28}H_{44}O$	

huge potential to aid understanding of the functionality of a protein. The 3D structure prediction of plipastatin synthase was done using prime module facilitated by Schrödinger Maestro suite. The structure prediction starts with obtaining the amino acid sequence from the protein database available at the website of NCBI. This is known as query sequence. The primary sequence of plipastatin synthase in FASTA format is fed into prime software to start the initial step of sequence alignment. The software uses ClustalW, a global alignment algorithm for searching the regions of similarity between the sequences. Sequence alignment resulted in the identification of the most similar sequence with the similarity score of 30.20% with PDB ID 2VSQ. This is also a NRPS that was used as a template to predict the backbone of the query sequence. Further addition of side chains, hydrogen and appropriate charge was performed to prepare the predicted model of plipastatin synthase. The model was further validated using a web-based tool named PROCHECK and a Ramchandran report is generated. Figure 5.2 shows the structure of plipastatin synthase predicted by homology modeling.

5.1.5 Molecular Docking of the Generated Model

Molecular docking for plipastatin was performed using the glide application of Maestro suite of Schrödinger software. Docking studies were performed to elucidate the interaction of drug plipastatin at its potential site of action (leucine and ergosterol) on fungal cells. The results of this interaction are shown in Table 5.2. The process of molecular docking starts with the preparation of lipopeptide plipastatin synthetase to optimize its structure. This step incorporates the removal

FIGURE 5.2   3D structure of plipastatin predicted by homology modeling.

TABLE 5.2   Docking Results of Plipastatin with Its Ligands

Ligands	Ergosterol	Leu
H Bond	1	4
Docking Score	−4.673	−3.365
Glide Energy (kcal/mol)	−30.648	−20.655

of water molecules and restrain minimization of the molecule using the protein preparation wizard in the application tab of Maestro. This step generates the maximum possible orientations of protein structure. Further, suitable sites are generated using sitemap application to perform docking. Ligands are docked against these generated sites to visualize this interaction in the binding pocket. Figure 5.3 represents the interaction of plipastatin with its target site ergosterol with the formation of one hydrogen bond. The docking score and glide energy is −4.673 and −30.648 kcal/mol, respectively. Figure 5.4 depicts the binding pocket of plipastatin for the ligand leu. The interaction of ligand leu and antifungal drug plipastatin is established by the formation of four hydrogen bonds and docking score and glide energy is −3.365 and −20.655 kcal/mol, respectively.

5.1.6 Pharmacokinetics of Plipastatin

Pharmacokinetic properties of drugs are important parameters to understand and decipher the metabolism and distribution of drug in a body. It is generally the prediction of ADMET properties of a drug, which stands for absorption, distribution, metabolism, excretion and

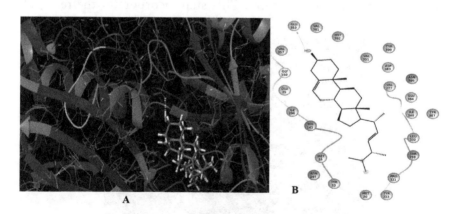

FIGURE 5.3   (A) Hydrogen bond interaction diagram, (B) ligand interaction diagram of plipastatin and ergosterol.

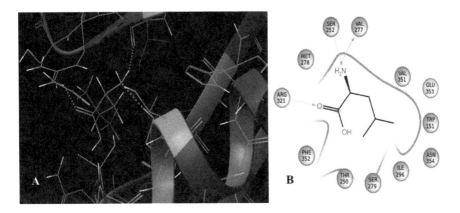

FIGURE 5.4 (A) Hydrogen bond interaction diagram, (B) ligand interaction diagram of plipastatin and leucine.

toxicity of a drug. This also reveals the environmental hazards of a drug after excretion. The pharmacokinetic properties of plipastatin are predicted through the web-based tool admetSAR. Results are displayed in Table 5.3. As per the data generated for absorption profiling, plipastatin is unable to cross the blood–brain barrier (BBB) as negative probability is given as 0.9927. Human intestinal absorption (HIA) is positive but the lower absorption value of 0.5716 predicts its lesser absorption probability. The negative probability indicates non Caco2 permeability. Plipastatin acts as a substrate for P-glycoprotein and hence it is a non-inhibitor of P-glycoprotein and does not interfere with its function. Plipastatin is also a non-inhibitor of renal organic cation transporter. The compound is subcellularly localized in lysosome. The metabolism profiling of plipastatin states that it is both a non-inhibitor and a non-substrate for the CYP450 family of enzymes except for CYP450 3A4 where it acts a substrate. The prediction of toxicity profiling of plipastatin is explained as non-carcinogenic and non-AMES toxic. The drug is predicted to be highly FHMT and has high TPT (*tetrahymena pyriformis* toxicity) and low HBT (honey bee toxicity) for fish, *Tetrahymena pyriformis* and honey bee toxicity profiling, respectively. The biodegradability profile also shows that it is not readily biodegradable. The acute oral toxicity is predicted as class III which is considered as harmful. The regression analysis for an ADMET predicted profile gives the results for solubility index (logS) as −3.2679, lethal dose LD50 value for rat acute toxicity as 3.2233 mol/kg and LC50 and GC50 values as 1.3798 mg/L and 0.4597 μg/L, respectively.

TABLE 5.3   Pharmacokinetics of Plipastatin

ADMET Predicted Profile – Classification		
Model	**Result**	**Probability**
Absorption		
Blood–Brain Barrier	BBB-	0.9927
Human Intestinal Absorption	HIA+	0.5716
Caco-2 Permeability	Caco2-	0.8260
P-glycoprotein Substrate	Substrate	0.8419
P-glycoprotein Inhibitor	Non-inhibitor	0.8917
	Non-inhibitor	0.8705
Renal Organic Cation Transporter	Non-inhibitor	0.9215
Distribution		
Subcellular localization	Lysosome	0.4960
Metabolism		
CYP450 2C9 Substrate	Non-substrate	0.8593
CYP450 2D6 Substrate	Non-substrate	0.8261
CYP450 3A4 Substrate	Substrate	0.5524
CYP450 1A2 Inhibitor	Non-inhibitor	0.9556
CYP450 2C9 Inhibitor	Non-inhibitor	0.9199
CYP450 2D6 Inhibitor	Non-inhibitor	0.8267
CYP450 2C19 Inhibitor	Non-inhibitor	0.8872
CYP450 3A4 Inhibitor	Non-inhibitor	0.7396
CYP Inhibitory Promiscuity	Low CYP Inhibitory Promiscuity	0.9318
Excretion		
Toxicity		
Human Ether-a-go-go-Related Gene Inhibition	Weak Inhibitor	0.9340
	Non-inhibitor	0.5998
AMES Toxicity	Non-AMES toxic	0.7732
Carcinogens	Non-carcinogens	0.8368
Fish Toxicity	High FHMT	0.9823
Tetrahymena pyriformis Toxicity	High TPT	0.9883
Honey Bee Toxicity	Low HBT	0.7399
Biodegradation	Not readily biodegradable	0.9928
Acute Oral Toxicity	III	0.5685
Carcinogenicity (Three-class)	Non-required	0.5920
ADMET Predicted Profile – Regression		
Model	**Value**	**Unit**
Absorption		
Aqueous Solubility	−3.2679	LogS
Caco-2 Permeability	−0.2837	LogPapp, cm/s
		(Continued)

TABLE 5.3 (CONTINUED)   Pharmacokinetics of Plipastatin

	Distribution Metabolism Excretion Toxicity	
Rat Acute Toxicity	3.2233	LD50, mol/kg
Fish Toxicity	1.3798	pLC50, mg/L
Tetrahymena pyriformis Toxicity	0.4597	pIGC50, µg/L

5.2 FUSARICIDIN SYNTHASE

5.2.1 Introduction

Fusaricidin is a depsipeptide antibiotic, which consists of six amino acids. It is produced by *Paenibacillus polymyxa* which was formerly known as *Bacillus polymyxa*. Fusaricidin is known to possess antifungal properties. Lipopeptide fusaricidins are generally a group of antibiotics that have a strong inhibitory role against the growth of various pathogenic fungi of plants. They consist of a cyclic hexapeptide linked with guanidinylated β-hydroxy fatty acid. The hexapeptide contains four amino acids of D-configuration (Nakajima et al. 1972, Kurusa et al. 1987, Kajimura and Kaneda 1996, Kajimura and Kaneda 1997, Kuroda et al. 2000). Fusaricidin is produced by a precursor *fusA* gene, a peptide synthetase gene. The fusaricidin synthetase gene is identified from the genome sequence of *Penibacillus polymyxa* E681, a rhizobacterium. The *fusA* gene encodes for a polypeptide that contains six modules in a single open reading frame. This gene also codes for a non-ribosomal peptide synthase (NRPS), fusaricidin synthase. (Li et al. 2007).

5.2.2 Synthesis of Fusaricidin from Fusaricidin Synthase

NRPSs are big multienzyme complexes which are organized into various modules (Marahiel et al. 1997). These modules are usually colinearly numbered and ordered with sequence of amino acids of peptide. These modules are further divided into corresponding domains, responsible for catalyzing each step of peptide chain synthesis. The first domain is adenylation domain (A) which takes part in substrate recognition and is responsible for activation of substrate as aminoacyl adenylate. Subsequently, activated amino acid is transferred to cofactor 4′phosphopentatheine (4′PP) which is tethered covalently to T domain. The T domain is located downstream to the A domain. Next comes the condensation domain (C), which is located between the A and T domains of successive domains and catalyzes the formation of peptide bond between two neighboring substrates. Ultimately, the release of fully assembled peptide chain from the template of enzyme takes place through hydrolysis or cyclization. Usually, this is carried out by the TE domain (thioesterase) which is located at the C-terminal end of the last module. Though, in some cases an additional reductase domain can also be found responsible for release and cyclization (Kessler et al. 2004, Koop et al. 2006). An interesting fact is that, the epimerization domain is absent in module six of gene *fusA*. This suggests that the amino acid at the sixth position at the fusaricidin analog produced by *P. polymyxa*, may be found to be an L-form. However, all fusaricidins are reported to have D-form of alanine residue at their sixth position (Kajimura and Kaneda 1996). It is reported that inactivation of the *fusA* gene leads to the loss of antifungal property against a plant fungal pathogen *Fusarium oxysporum*. The LC/MS analysis of *fusA* mutant strain shows its inability to produce fusaricidin which confirms the involvement of *fusA* for fusaricidin biosynthesis. As per the findings, *fusA* is capable of producing various forms of fusaricidin, which has been identified from *P. polymyxa* E681 (Choi et al. 2008). Fusaricidin A is a newly identified depsipeptide antibiotic consisting of a unique fatty acid, that is, 15-guanidino-3-hydroxypentadecanoic acid. It was isolated from *Bacillus polymyxa* KT-8 culture broth along with Fusaricidin B, C and D as minor constituents. The rhizosphere of garlic suffering from basal rot disease and infected with *Fusarium oxysporum* was used to obtain fusaricidin A producing *B. polymyxa* KT-8. The two-dimensional structure of fusaricidin is shown in Figure 5.5 which was drawn using the 2D sketcher tool of the Maestro suite of Schrödinger.

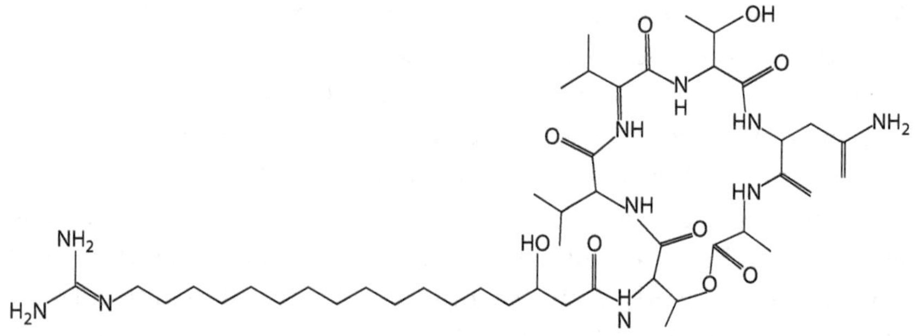

FIGURE 5.5　2D structure of fusaricidin.

5.2.3 Cytotoxic Effect of Fusaricidin Lipopeptide

Antimicrobial lipopeptides are a cluster of small peptides linked with fatty acids which are secreted by various microorganisms. These represent a new promising class of antibiotics. These are found to be active against various ranges of microorganisms which include fungi, bacteria, virus, protozoa, yeast and tumor cells as well. Such active lipopeptides have low molecular weight, high stability, high efficiency, lesser drug resistance and a specific mechanism of action. Fusaricidin, a lipopeptide antibiotic with six amino acids is identified as an antifungal lipopeptide from *P. polymyxa*. Fusaricidin A is found to be potent against fungi as well as Gram-positive bacteria while fusaricidin C and D possess strong activity specifically against Gram-positive bacteria, particularly, *Staphylococcus aureus, S. aureus* FDA 209P and *Micrococcus leuteus* IFO3333, similar to fusaricidin A (Kajimura and Kaneda 1996). However, fusaricidin B possesses a weaker potency against the previously mentioned microbes compared to the mixture of fusaricidin C and D. Fusaricidin shows no activity against Gram-negative bacteria tested even at 100 µg/mL (Kajimura and Kaneda 1996, Kajimura and Kaneda 1997, Yu et al. 2012).

The cytotoxic mechanism of action lies in its activity on cell membranes. According to the experimental data available, the addition of fusaricidin to the culture of *Bacillus subtilis* leads to destruction of the cell membrane and increased production of hydroxyl ions (OH^-). This, at the initial level, affects the biosynthesis of nucleic acid and protein of cells. The novel antibiotics should stimulate the cells to secrete more OH^- to disturb the growth of the cells and simultaneously prevent them from congesting (Yu et al. 2012). The transcriptome analysis of fusaricidin reveals that it induces a set of genes by exposure of membrane-active compounds. The genetic studies indicated the sensitivity of Sigma factor (SigA) to this change. The consistency of these notions resulted in the conclusion that such type of antibiotics primarily acts on the cell membrane (Hachmann et al. 2009). Apparently, *B. subtilis* is found to have the ability to alter its gene expression to develop resistance in response to fusaricidin exposure and some other environment changes.

According to a phenomenon, the treatment of fusaricidin increases the catabolism of protein and fatty acids but has a strong ability to repress the decomposition of glucose and gluconeogenesis. This indicates that more energy is required by microbial strains to support defense against peptide-based antibiotics which suggests the insight of developing novel antibiotic drugs. The Munich Information Center for Protein Sequences (MIPS) analysis for genome annotation also revealed the significant alteration in the genes involving metabolism of nucleotides. The pathway analysis suggests that purine and pyrimidines synthesis is repressed at an early stage of fusaricidin treatment. The increase in degradation of nucleotide precursor by fusaricidin treatment indicates reduction in the availability of nucleic acid constituents in *B. subtilis* because of the antibiotic (Yu et al. 2012).

5.2.4 Identification of Ligands

Ligand identification is performed to simulate the molecular docking analysis of fusaricidin. The mechanism of action indicates that the inhibitory action of fusaricidin starts with the activity of antibiotic on the cell membrane. The perturbation on the cell membrane is facilitated by the production of OH^- which further alters the response of a certain set of genes in the cell to hinder the cellular biosynthesis activity. To understand the membrane action of fusaricidin, ligands participating on the cell surface need to be identified. First, the structure prediction of fusaricidin was done through homology modeling through the software-based prediction by Prime module facilitated by Schrödinger Maestro suite. The amino acid sequence of fusaricidin was obtained through the protein database available at NCBI. The FASTA sequence was further subjected to sequence alignment through ClustalW, a global alignment program used by Prime. The sequence alignment resulted in the template protein 2VSQ, which is a NRPS. The model of fusaricidin was built on template protein 2VSQ. The unique ligand of 2VSQ, i.e. leucine was identified through the RCSB website. The leucine molecule was drawn using the 2D sketcher tool on the Maestro application to obtain its 3D structure. Ligand data with its chemical formula and 2D structure is represented in Table 5.4.

5.2.5 Ligand-Mediated Molecular Docking

Molecular docking is a very powerful approach to interpret the interaction of two molecules. Generally, it is performed to visualize the

TABLE 5.4   Ligands for Fusaricidin

Ligand	IUPAC Name & Molecular Formula	2D Structure
Leu	Leucine $C_6H_{13}NO_2$	

interaction of a drug and a target. The molecular docking process can be concluded with the interaction through the number of bonds formed between the drug and target. This gives a docking score and energy of the complex in kcal/mol, which is used to understand the interaction. Hence it is considered to be an essential step in drug discovery. Various licensed and freely available software are used to perform *in silico* docking. Thus, it is also referred to as computer-aided drug discovery. In this current study, we used Schrödinger software to perform ligand-mediated molecular docking. The software provides a Maestro application for computer-aided drug design and discovery. The application facilitates structure prediction of the drug and the ligand molecule though various modules like prime and 2D sketcher. The molecular docking process is performed using the glide module. To start the docking process, first, protein molecule is prepared using the protein preparation wizard. This step helps to optimize the structure and minimize energy of protein molecule. The step is followed by the identification of the docking site using the sitemap application with the generation of the possible binding pockets. Further grid generation through glide is followed by molecular docking. Molecular docking of fusaricidin with ligand leucine gave the docking score of −4.384 and glide energy of −31.729 kcal/mol. The interaction is mediated by the formation of four hydrogen bonds between the lipopeptide and the ligand. The data suggests the considerable interaction between the lipopeptide-based drug fusaricidin and the proposed cell membrane ligand leucine to perform the antibiotic action as a possible drug candidate. The docking results are summarized in Table 5.5. Figure 5.6(A) depicts the hydrogen bond interaction diagram between the fusaricidin and leucine while Figure 5.6(B) shows the ligand interaction diagram between the participating amino acids in the binding pocket of fusaricidin. According to the diagram, the amino and hydroxyl group of leucine interacts with glycine at 398, leucine at 478 and glutamic acid at 293 positions of fusaricidin.

TABLE 5.5   Docking Results of Fusaricidin with Its Ligand

Ligands	Leu
H Bond	4
Docking Score	−4.384
Glide Energy (kcal/mol)	−31.729

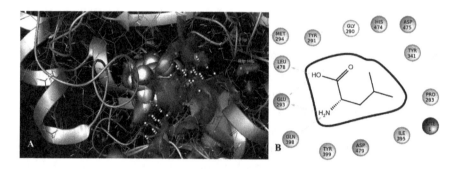

FIGURE 5.6 (A) Hydrogen bond interaction diagram, (B) ligand interaction diagram of fusaricidin and leucine.

5.2.6 Drug Behavior Studies Using ADMET

Drug behavior and drug likeness of a proposed drug molecule can be predicted through ADMET properties *in silico*. These properties include absorption, distribution, metabolism, excretion and toxicity analysis of a candidate drug. In computer-aided drug design, prediction of ADMET properties is an essential step to understand the pharmacokinetic properties of a drug molecule. Various computational tools and algorithms are developed on the basis of previously available data of molecular descriptors which helps to predict structure activity relationships. This ultimately compiles the prediction of behavior of drug and drug likeness on various parameters. In this study, we have used a web-based tool named admetSAR to predict the ADMET properties of fusaricidin. The canonical SMILES were obtained from the PubChem compound database available at the NCBI website. The predicted results have been categorized on the basis of fate of drug during consumption and metabolism. These parameters help in deciphering the mode of administration of drug on the basis of absorption and metabolism profiling. The results are displayed in Table 5.6. The absorption profiling predicted the BBB crossing ability as negative with the probability of 0.9686. This ensures the inability of drugs to cross the BBB. HIA negative with probability index 0.8404 deciphers the inability of fusaricidin for intestinal absorption. This is because of the high molecular weight of the drug, which gave BBB and HIA negative results. Further negative Caco2 permeability is predicted. PgP and PgS profiling suggested it is a non-inhibitor and substrate, respectively depicting the non-interference of PgP transport for drug efflux of xenobiotic compounds. Non-inhibition of renal organic cation transporter deciphers the drug to be non-interfering with

TABLE 5.6 Pharmacokinetics of Fusaricidin

ADMET Predicted Profile – Classification		
Model	**Result**	**Probability**
	Absorption	
Blood–Brain Barrier	BBB-	0.9686
Human Intestinal Absorption	HIA-	0.8404
Caco-2 Permeability	Caco2-	0.7352
P-glycoprotein Substrate	Substrate	0.7367
P-glycoprotein Inhibitor	Non-inhibitor	0.7941
	Non-inhibitor	0.9186
Renal Organic Cation Transporter	Non-inhibitor	0.9023
	Distribution	
Subcellular Localization	Mitochondria	0.6449
	Metabolism	
CYP450 2C9 Substrate	Non-substrate	0.8352
CYP450 2D6 Substrate	Non-substrate	0.8073
CYP450 3A4 Substrate	Substrate	0.5090
CYP450 1A2 Inhibitor	Non-inhibitor	0.9104
CYP450 2C9 Inhibitor	Non-inhibitor	0.9231
CYP450 2D6 Inhibitor	Non-inhibitor	0.9277
Vxz CYP450 2C19 Inhibitor	Non-inhibitor	0.9113
CYP450 3A4 Inhibitor	Non-inhibitor	0.9502
CYP Inhibitory Promiscuity	Low CYP Inhibitory Promiscuity	0.9959
	Excretion	
	Toxicity	
Human Ether-a-go-go-Related Gene Inhibition	Weak inhibitor	0.9709
	Non-inhibitor	0.8927
AMES Toxicity	Non-AMES toxic	0.6930
Carcinogens	Non-carcinogens	0.9365
Fish Toxicity	Low FHMT	0.7148
Tetrahymena pyriformis Toxicity	High TPT	0.9147
Honey Bee Toxicity	Low HBT	0.7380
Biodegradation	Not readily biodegradable	0.8977
Acute Oral Toxicity	III	0.5990
Carcinogenicity (Three-class)	Non-required	0.6073
ADMET Predicted Profile – Regression		
Model	**Value**	**Unit**
	Absorption	
Aqueous Solubility	−2.5208	LogS
Caco-2 Permeability	−0.5466	LogPapp, cm/s
		(*Continued*)

TABLE 5.6 (CONTINUED)   Pharmacokinetics of Fusaricidin

	Distribution Metabolism Excretion Toxicity	
Rat Acute Toxicity	2.7458	LD50, mol/kg
Fish Toxicity	1.9396	pLC50, mg/L
Tetrahymena pyriformis Toxicity	0.1156	pIGC50, µg/L

kidney functions with a good probability index. Distribution profiling is predicted to be subcellular localization in mitochondria. The metabolism profile is predicted for CYP450 enzymes as non-substrate and non-inhibitor with probability of 80–90% except as substrate for CYP450 3A4 substrate with 50% probability. Low inhibitory promiscuity for CYP inhibition is predicted with 0.99 probability. The toxicity profiling on various measures has been depicted individually in table with no AMES toxicity, no carcinogenic, low FHMT, high TPT, low HBT and as not readily available for biodegradation. The oral acute toxicity is class III with 50% probability. This suggests the harm in oral consumption of the drug. Further regression analysis predicts logS (aqueous solubility) and logP as −2.5208 and −0.5466, respectively with LD50 (lethal dose for rat acute toxicity), pLC50 (Fish toxicity) and pIGC50 as 2.7458 mol/kg, 1.9396 mg/L and 0.1156 µg/L, respectively.

5.3 SUMMARY

Lipopeptide synthases play an important role in the synthesis of lipopeptides. They are multienzyme complexes which have various modules for peptide biosynthesis. The structures of two lipopeptides synthetases, plipastatin synthase and fusaricidin synthase, were obtained by homology modeling on a NRPS with the corresponding lipopeptides. Once the structures of the synthases were obtained, they were docked with the various ligands to obtain the ligand and their pharmacokinetic properties after uptake of drug. The results demonstrate that among the two ligands for plipastatin synthase, leucine and ergosterol, ergosterol exhibits better docking with a docking score of −4.673 and glide energy of −30.648 kcal/mol. On the other hand, only one ligand, leucine, is suitable for fusaricidin synthase and had a docking score of −4.384 with −31.729 kcal/mol as the glide energy.

REFERENCES

Choi, S.-K., Park, S.-Y., Kim, R., et al. 2008. Identification and functional analysis of the fusaricidin biosynthetic gene of *Paenibacillus polymyxa* E681. *Biochemical and Biophysical Research Communications* 365: 89–95.

Deleu, M., Paquot, M. and Nylander, T. 2005. Fengycin interaction with lipid monolayers at the air–aqueous interface—implications for the effect of fengycin on biological membranes. *Journal of Colloid and Interface Science* 283: 358–365.

Falardeau, J., Wise, C., Novitsky, L. and Avis, T. 2013. Ecological and mechanistic insights into the direct and indirect antimicrobial properties of *Bacillus subtilis* lipopeptides on plant pathogens. *Journal of Chemical Ecology* 39: 869–878.

Gong, A.-D., Li, H.-P., Yuan, Q.-S., et al. 2015. Antagonistic mechanism of iturin A and plipastatin A from *Bacillus amyloliquefaciens* S76-3 from wheat spikes against *Fusarium graminearum*. *PloS One* 10: e0116871.

Hachmann, A.-B., Angert, E. R. and Helmann, J. D. 2009. Genetic analysis of factors affecting susceptibility of *Bacillus subtilis* to daptomycin. *Antimicrobial Agents and Chemotherapy* 53: 1598–1609.

Honma, M., Tanaka, K., Konno, K., et al. 2012. Termination of the structural confusion between plipastatin A1 and fengycin IX. *Bioorganic &Medicinal Chemistry* 20: 3793–3798.

Horn, J. N., Cravens, A. and Grossfield, A. 2013. Interactions between fengycin and model bilayers quantified by coarse-grained molecular dynamics. *Biophysical Journal* 105: 1612–1623.

Kajimura, Y. and Kaneda, M. 1996. Fusaricidin A, a new depsipeptide antibiotic produced by *Bacillus polymyxa* KT-8. *The Journal of Antibiotics* 49: 129–135.

Kajimura, Y. and Kaneda, M. 1997. Fusaricidins B, C and D, new depsipeptide antibiotics produced by *Bacillus polymyxa* KT-8: isolation, structure elucidation and biological activity. *The Journal of Antibiotics* 50: 220–228.

Kessler, N., Schuhmann, H., Morneweg, S., Linne, U. and Marahiel, M. A. 2004. The linear pentadecapeptide gramicidin is assembled by four multimodular nonribosomal peptide synthetases that comprise 16 modules with 56 catalytic domains. *Journal of Biological Chemistry* 279: 7413–7419.

Koop, F., Mahlert, C., Grünewald, J. and Marahiel, M. A. 2006. Peptide macrocyclization: the reductase of the nostocyclopeptide synthetase triggers the self-assembly of a macrocyclic imine. *Journal of the American Chemical Society* 128: 16478–16479.

Kuroda, J., Toshio, F. and Kionishi, M. 2000. LI-F antibiotics, a family of antifungal cyclic depsipeptides produced by *Bacillus polymyxa* L-1129. *Heterocycles* 53: 1533–1549.

Kurusa, K., Ohba, T., Arai, T. and Fukushima, K. 1987. New peptide antibiotics LI-F03, F04, F05, F07, and F08, produced by *Bacillus polymyxa*. I. Isolation and characterization. *The Journal of Antibiotics* 40: 1506–1514.

Li, J., Beatty, P. K., Shah, S. and Jensen, S. E. 2007. Use of PCR-targeted muta-genesis to disrupt production of fusaricidin-type antifungal antibiotics in *Paenibacillus polymyxa*. *Applied and Environmental Microbiology* 73: 3480–3489.

Lin, T.-P., Chen, C.-L., Fu, H.-C., et al. 2005. Functional analysis of fengycin synthetase FenD. *Biochimica et Biophysica Acta (BBA)-Gene Structure and Expression* 1730: 159–164.

Maget-Dana, R. and Peypoux, F. 1994. Iturins, a special class of pore-forming lipopeptides: biological and physicochemical properties. *Toxicology* 87: 151–174.

Maget-Dana, R., Ptak, M., Peypoux, F. and Michel, G. 1985. Pore-forming prop-erties of iturin A, a lipopeptide antibiotic. *Biochimica et Biophysica Acta (BBA)-Biomembranes* 815: 405–409.

Marahiel, M. A., Stachelhaus, T. and Mootz, H. D. 1997. Modular peptide syn-thetases involved in nonribosomal peptide synthesis. *Chemical Reviews* 97: 2651–2674.

Nakajima, N., Chihara, S. and Koyama, Y. 1972. A new antibiotic, gatavalin. *The Journal of Antibiotics* 25: 243–247.

Nasir, M. N., Thawani, A., Kouzayha, A. and Besson, F. 2010. Interactions of the natural antimicrobial mycosubtilin with phospholipid membrane models. *Colloids and Surfaces B: Biointerfaces* 78: 17–23.

Ongena, M. and Jacques, P. 2008. *Bacillus* lipopeptides: versatile weapons for plant disease biocontrol. *Trends in Microbiology* 16: 115–125.

Ongena, M., Jacques, P., Touré, Y., et al. 2005. Involvement of fengycin-type lipo-peptides in the multifaceted biocontrol potential of *Bacillus subtilis*. *Applied Microbiology and Biotechnology* 69: 29.

Patel, H., Tscheka, C., Edwards, K., Karlsson, G. and Heerklotz, H. 2011. All-or-none membrane permeabilization by fengycin-type lipopeptides from *Bacillus subtilis* QST713. *Biochimica et Biophysica Acta (BBA)-Biomembranes* 1808: 2000–2008.

Raaijmakers, J. M., De Bruijn, I. and de Kock, M. J. 2006. Cyclic lipopeptide pro-duction by plant-associated *Pseudomonas* spp.: diversity, activity, biosyn-thesis, and regulation. *Molecular Plant-Microbe Interactions* 19: 699–710.

Steller, S., Vollenbroich, D., Leenders, F., et al. 1999. Structural and functional organization of the fengycin synthetase multienzyme system from *Bacillus subtilis* b213 and A1/3. *Chemistry &Biology* 6: 31–41.

Tosato, V., Albertini, A. M., Zotti, M., Sonda, S. and Bruschi, C. V. 1997. Sequence completion, identification and definition of the fengycin operon in *Bacillus subtilis* 168. *Microbiology* 143: 3443–3450.

Tsuge, K., Ano, T., Hirai, M., Nakamura, Y. and Shoda, M. 1999. The genes degQ, pps, and lpa-8 (sfp) are responsible for conversion of *Bacillus subtilis* 168 to plipastatin production. *Antimicrobial Agents and Chemotherapy* 43: 2183–2192.

Umezawa, H., Aoyagi, T., Nishikiori, T., et al. 1986. Plipastatins: New inhibi-tors of phospholipase A2, produced by *Bacillus cereus* BMG302-fF67. *The Journal of Antibiotics* 39: 737–744.

Vanittanakom, N., Loeffler, W., Koch, U. and Jung, G. 1986. Fengycin-a novel antifungal lipopeptide antibiotic produced by *Bacillus subtilis* F-29-3. *The Journal of Antibiotics* 39: 888–901.

Yu, W.-B., Yin, C.-Y., Zhou, Y. and Ye, B.-C. 2012. Prediction of the mechanism of action of fusaricidin on *Bacillus subtilis*. *PloS One* 7: e50003.

Index